The Farmer's EARTHWORM HANDBOOK

Managing Your Underground Money-Makers

By David Ernst

Lessiter Publications, Brookfield, Wis.

Publisher's Cataloging in Publication
(Prepared by Quality Books, Inc.)

Ernst, David T., 1958-
    The farmer's earthworm handbook : managing your underground
money-makers / David Ernst. — 1st ed.
    p. cm.
    Includes bibliographical references.
    ISBN 0-944079-03-2

    1. Zoology, economic. 2. Tillage. 3 Earthworms. I. Title.

SB998.E4E76 1995   591.6
                        QBI94-21151

International Standard Book
Number: 0-944079-03-2.

Published by Lessiter Publications,
P.O. Box 624, Brookfield, Wisconsin 53008-0624.

Manufactured in the United States of America.

Cover Design: Greg Kot

This book is dedicated to my grandfather,
Anthony Hollis Hemmingson,
who knew the value of a good fishing worm.

Table Of Contents

"The plow is one of the most ancient and most valuable of man's inventions; but long before he existed the land was plowed, and still continues to be plowed, by earthworms."

—*Charles Darwin, 1881*

NONE OF US is old enough to have personally interviewed Charles Darwin. But we have heard the gist of this famous quote on earthworms repeated again and again in conversations with farmers, in university research papers and articles in the farm press—nature has furnished an army of creatures that will enhance your soil's tilth and fertility if you let them.

For this book, we sifted through a mountain of material for the most valuable nuggets of earthworm wisdom. Our purpose was to explain how earthworm activity can make your farm more productive while conserving precious topsoil.

You will also find many valuable tips on how you can promote earthworm population growth without radically changing your current management, cropping or tillage practices.

Some of this information is drawn from thoroughly documented scientific data. On some pages, farmers relate their personal observations of earthworm behavior which are not necessarily less accurate than formal studies. A variety of philosophies are represented, but you will quickly notice what's widely accepted as truth despite the differing views.

Ultimately, we think you will find plenty of information in this book to help you farm better and much more profitably.

—*Dave Ernst*

Nature's Plow, Nature's Fertilizer Factory

MENTION THE NAME Charles Darwin and most people think of *Origin of Species,* a book that continues to stir up plenty of controversy today.

In some quarters, however, Darwin is remembered not so much as the grand theorist of evolution, but as an early and perceptive student of earthworms.

Like a growing number of scientists and farmers today, Darwin was intensely interested in earthworms and their impact on the soil. He summarized more than 40 years of observations in a book, *The Formation Of Vegetable Mould Through The Action Of Worms,* published a year before his death.

The book explained how earthworms help break down organic matter and maintain soil fertility. This idea was considered in its own way to be almost as radical as Darwin's thoughts on human origins, since many of Darwin's contemporaries believed earthworms were harmful to plants.

Changing Attitudes

Attitudes toward earthworms changed little for 60 years after Darwin's passing. As recently as 1943, one soil scientist dismissed earthworms by saying "Their direct effect on plant growth is minimal."

One spokesman for an alternative point of view was Edward Faulkner, author of *Plowman's Folly,* a pioneering book on conservation tillage published in 1943.

Faulkner noted one of the distinctive features of badly eroded soil was the lack of beneficial soil life such as earthworms. "Be-

TODAY'S PLOWING SYSTEM. *Various kinds of earthworms are replacing the moldboard plow and other tillage implements as the primary means of working the fields found on many of today's farms.*

cause of the dependence of these small life forms on decaying organic matter, the disappearance of the organic matter from our soils has caused a complete change in the fauna of the soil surface," Faulkner wrote. "The eviction of minute forms of life sets the stage for those large problems of drainage...The remedy is to restore at once the organic condition of the soil and with it the teeming life which feeds upon it."

Earthworm Excitement

Since then, however, agricul-

tural research and practical farming have revealed much more about the beneficial role earthworms play in improving soil structure and fertility. Farmer interest in earthworms has grown along with concern over excessive tillage and use of crop chemicals.

Some producers have switched to alternative tillage systems like no-till and ridge till or dispensed with application of commercial fertilizer and pesticides. Even "conventional" farmers agree that it may be best, economically and environmentally, to apply purchased inputs and work the soil only on an "as needed" basis.

If they provide a favorable environment for the tiny creatures, they have found that much seedbed preparation will be done for them. The more they learn about worms, the more excited they get.

Farmers Get Emotional

Several years ago scientists at DowElanco began studying how insecticides affect earthworms.

"It's been absolutely amazing to discover the significance of earthworms, especially in no-till situations, and the emotions farmers which have in regard to earthworms," says Ken Phelps, DowElanco product marketing manager.

"As a child, I remember following behind my father as he moldboard plowed in the spring," says Mark May, Wilton, Iowa. "The sweet smell of freshly turned soil was there and you could see earthworms—nature's own small-scale fertilizer factories.

"My brothers and I filled buckets to the brim with worms for fishing or just for the fun of it all. At the time, it only seemed natural that, as you turned the soil, earthworms would be there."

But, May adds, in recent years nature's link to soil regeneration has been overlooked. "The amount of earthworms present

---

FARMING PRACTICES IN ORDER OF EARTHWORM PREFERENCE

Tillage...
- ★ No-till
- ★ Ridge till
- ★ Chisel plow
- ★ Tandem disk
- ★ Moldboard plow
- ★ Offset disk

Mechanical Weed Control...
- ☞ Cultivation
- ☞ Rotary hoeing

Herbicides...
- ➡ Banded
- ➡ Surface broadcast
- ➡ Disk incorporated

Fertilizer...
- ➡ Lightly bulked manure
- ➡ Rotated perennials
- ➡ Slurry
- ➡ High ammonia

—Leopold Center for Sustainable Agriculture, Ames, Iowa

---

are the best indicator of the life of the soil," he says. "That's why I was ecstatic to find our ridge tilled fields covered with castings

SUPER SEEDBED. With worms doing much of the tillage in conservation tillage, farmers have a higher quality seedbed at planting time.

of earthworms."

The feeding, casting and burrowing action of earthworms is like plowing in the way it stirs the soil, mixing crop residue, air and water into the soil profile, says Eileen Kladivko, Purdue Univ. soil management specialist.

"We may be able to replace some of the mechanical energy inputs for conventional tillage with the increased activity of biological populations," she says.

Worms Vs. Fertilizer?

Bill Becker, crop consultant and research director for the Central Illinois Research Farm near Springfield, Ill., says increased earthworm activity can reduce the need for commercial fertilizer. Potential savings for farmers are in the $20,000 to $30,000 per year range for a typical farm.

The specific advantages of earthworm activity include:

1 Improved Water Infiltration:

The pencil-thin burrows which earthworms drill allow rainwater to penetrate deep into the subsoil.

Gentle slopes under a July crop canopy can absorb a 4-in. downpour in 2 hours with virtually no runoff.

In many soils, worms will maintain their burrows even after crop cultivation operations are completed.

Because water seeps slowly from the burrows into the surrounding soil, the entire root zone is used for long-term moisture storage. The earthworm burrows help avoid waterlogged soils by speeding excess water to tile lines.

2 Improved Soil Aeration:

Vertical burrows also pipe air from the soil surface down to a depth of 5-ft. The oxygen stimulates microbial conversion of minerals into plant nutrients. Carbon dioxide produced by soil life is carried up to the surface where it stimulates plant leaf growth.

3 Increased Hardpan Penetration:

Earthworm tunnels remain in place for several years because the worms coat them with sticky, nutrient-rich mucus. Oxygen, nutrients and bacterial life encourage plant roots to grow down, perforating the plow pan.

As generations of root residue fill the smaller tunnels, deep moisture moves up by capillary action to roots nearer the surface.

4 Reduced Soil Compaction:

Horizontal burrows often allow roots to grow 6 to 7-in. a day through otherwise compacted soil.

In 1986, Dale McNelly, an Arcanum, Ohio, ridge tiller told Ohio State Univ. farm management specialist Don Moore that soil drainage had improved since he switched to ridge till 5 years before.

At first he thought the improvement resulted solely from reduced compaction through controlled field traffic. After reassessing his situation, however, he concluded earthworm burrows improved drainage even before compaction was eliminated.

5 Surface Residue Mixed Into Soil:

Some species of worms pull crop residue into their underground passages and, under good conditions, are capable of incorporating the entire year's production of residue. Once underground, they eat and digest the

---

"We filled buckets to the brim with worms for fishing. It only seemed natural that as you turned the soil, earthworms would be there..."
—Mark May, Wilton, Iowa

---

raw residue, converting it to nutrient-rich humus and other compounds beneficial to crops.

Their droppings are called castings and 63,000 worms in an acre of good soil deposit 18 tons of castings each year. In 10 years' time, this may completely renew the top 2-in. of topsoil.

6 Release Of Crop Growth Stimulants:

Earthworms break down crop growth inhibitors, such as phenols and formaldehyde, from decaying residue and lace the processed material with growth stimulants containing auxins and cytokinins.

In one year, they can do this for 50 tons of soil per acre or about 5% of the total weight of 6-in. of topsoil. In 20 years, they can process the entire topsoil layer.

7 Minerals Brought Up From Subsoil:

Earthworm deposits contain 5 to 10 times the amount of soluble plant nutrients as the original soil. This action alone may increase soil test results for calcium, nitrogen, phosphorus and potassium.

8 Neutralization Of Soil pH:

Soil processed by earthworms is always closer to neutral pH than the original, whether acid or alkaline.

9 Rapid Growth Of

Beneficial Microbes:

The earthworm carries microbes in its body. After they're deposited along burrows and on the soil surface, these microbes help decompose crop residues.

10 Improved Soil Tilth:

Earthworm-treated soil contains sugars and enzymes which stabilize the soil particles into clumps or aggregates. This improves infiltration and water-holding capacity.

A Purdue Univ. greenhouse study showed soil infiltration rates increased 8 to 15-fold when 16-liter pots contained 15 and 30 earthworms respectively. Soil crusting from simulated rainfall was greatly reduced by earthworm activity.

South Dakota State Univ. research indicates fields with high earthworm populations have infiltration rates four times higher than comparable fields without earthworms.

Jeff Duncan of Leesburg, Ohio, recalls when the local electric company came to lay a cable across his land after he had been no-tilling for 10 years. The cable also went across the property of a conventional tilling neighbor.

"The bulldozer and plow the company was using could operate in a higher gear on my farm than on my neighbor's," Duncan says. He credits the improved soil tilth to increased earthworm activity.

11 Improved Nematode Control:

Earthworms eat harmful nematodes and create soil conditions which discourage increases in nematode numbers.

12 Increased Micronutrient Chelation:

Earthworms link micronutrients, such as zinc and boron, with other nutrients so plant roots absorb them more readily.

Put Worms To Work!

This book will show you how to immediately boost earthworm activity regardless of your current tillage and cropping practices. It may involve simple decisions about which products to use and when to apply them.

You'll learn whether it's worth your while to switch to no-till, ridge till or other conservation tillage systems—and how earthworms can help you succeed.

You can flip ahead to the chapter which most interests you or gain valuable background for your decision-making by reading the whole book. Whatever way best suits you, you can take advantage of one of nature's wonders—the earthworm!

Sources

Biology Of Earthworms, C.A. Edwards and J.R. Lofty, Chapman and Hall, 1977.

Earthworms, Their Ecology And Relationships With Soils And Land Use, K.E. Lee, Academic Press, 1985.

Faunal Indicators Of Soil Quality, D.R. Linden, P.F. Hendrix, D.C. Coleman and P.C.J. van Vliet, USDA Agricultural Research Service, St. Paul, Minn., and Univ. of Georgia, Athens, Ga., January, 1993.

"Hooked On Worms," Darrell Smith, *Farm Journal,* January, 1993.

"Interaction Of Earthworms With Soil Physical Conditions Influencing Plant Growth," S.D. Logsdon and D.R. Linden, *Soil Science,* October, 1992, Vol. 154, No. 4.

"Land's Best Friend," *LandOwner,* Cedar Falls, Iowa (reprint).

"Micromorphic Observation Of Soil Alteration By Earthworms," L.T. West, P.F. Hendrix and R.R. Bruce, *Agriculture, Ecosystems And Environment,* Elsevier Science Publishers B.V., Amsterdam, 1991.

"Microrelief And Rainfall Effects On Water And Solute Movement In Earthworm Burrows," M.D. Trojan and D.R. Linden, *Soil Science Of America Journal,* May-June, 1992, Vol. 56, No. 3.

"No-Till Keeps Farm From Slipping Away," *Ohio Farmer,* April, 1992.

The Formation Of Vegetable Mould Through The Action Of Worms, Charles Darwin, reprinted in 1972 by AMS, 56 E. 13th St., New York, NY

> *"My earthworm population has virtually exploded since I quit killing them with tillage..."*
> —Randy Henke, Sawyer, N.D.

10003-4686.

Plowman's Folly, Edward H. Faulkner, Grosset & Dunlap, 1943.

"Retailers Discuss Lorsban 15G With DowElanco Manager," *Ag Retailer*, Oct. 1993.

Soil Ecology, Ken Killham, Cambridge University Press, 1994.

"What Are Worms Good For?", *Dakota Farmer*, August 17, 1991.

"Worming Around," *American Farmland*, Spring, 1993.

EARTHWORM ARITHMETIC...

25 earthworms per sq. ft. = 1 ton of worms per acre.

Annual work schedule for earthworms = 100 days a year.

1 ton worms = 100 tons of castings or 2/3-in. manure on surface per acre.

Macropore equivalent = 4,000-ft. of 6-in. tile per acre.

Tiling at $1.20 per ft. = $4,800 per acre.

Nutrients which worms typically add to an acre of soil each year:

★ 4-lbs. nitrate of nitrogen.

★ 30-lbs. of phosphorus.

★ 72-lbs. of potash.

★ 90-lbs. of magnesium.

★ 500-lbs. of calcium.

Analysis = 4-68-96 and 3/4-ton of limestone.

Value = $34.15 per acre.

—Bill Becker, crop consultant, Springfield, Ill.

The Basics Of Earthworm Biology

THE TEXAS BORDER TOWN of Laredo was gripped by earthworm mania in March, 1993, after the local newspaper reported that a 79-ft.-long, 300-lb. earthworm was leaving a sticky trail of mucus along Interstate 35. Curiosity seekers flocked to the local Target store parking lot where authorities supposedly kept the captured monster.

Of course, it was a hoax. A reporter (who was soon out of a job) wrote the supermarket tabloid style story in an idle moment and it inexplicably found its way into print. But the episode illustrates how little people know about a simple but vitally important animal.

Spineless Wonders

According to reliable records, the very largest earthworms in the world measure 3 to 5-ft. in length, about 1-in. in diameter and weigh a maximum of 1.3-lbs. These big ones are found mostly in the Southern Hemisphere.

Like all invertebrates, earthworms are limited in size because they have no backbones to support their weight. Unlike most invertebrates, they have a closed system of blood vessels with muscular valves that function as "hearts."

The essential earthworm consists of two concentric tubes, the body wall and the gut, separated by a fluid-filled cavity, the coelom. Its body is divided into segments by thin membranes called septa.

To move from place to place, the earthworm contracts and relaxes its body segments in sequence. An earthworm can move up a sheet of polished metal or clean glass, using its mouth as a suction cup if necessary. Most of the time it uses bristly organs called setae to grip slippery surfaces.

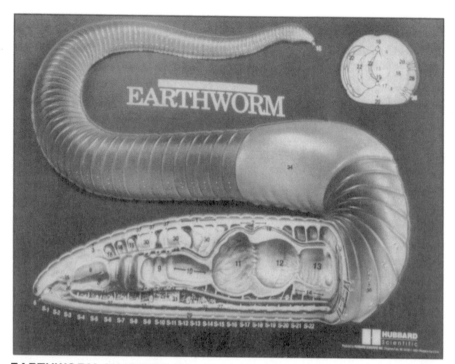

EARTHWORM ANATOMY. *The Hubbard Earthworm Model shows the internal and external anatomy of the common earthworm, Lumbricus.*
The supplemental diagram at upper right relates the mid-dorsal section of the earthworm in its lateral view to the body plan as shown in the cross-section drawing. This particular model is unique in that it shows the entire earthworm from the prostomium to the anus.

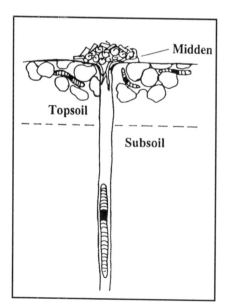

ENVIRONMENT CROSS-SECTION. *Many types of earthworms feed on organic matter found on the soil surface and in the first few inches of topsoil. This residue is often piled into middens which are piles of residue stacked on top of worm holes in the field. These middens are readily noticeable as you walk through your fields.*

KEYS TO EARTHWORM MODEL SHOWN AT LEFT...

1. Circular muscles.*
2. Longitudinal muscles.
3. Cuticle and epidermis.
4. Coelom.
5. Prostomium.
6. Mouth.
7. Buccal cavity.
8. Pharynx.
9. Esophagus.
10. Calciferious glands.
11. Crop.
12. Gizzard.
13. Intestine.
14. Chloragogen cells.
15. Anus.
16. Nephridium—bladder and excretory tubules.
17. Nephridium—funnel.
18. Dorsal vessel.
19. Ventral vessel.
20. Subneural vessel.
21. Parietal vessel.
22. Blood vessels to intestine, nephridia and body wall.
23. Hearts—two of five.
24. Three hearts cut away.
25. Suprapharyngeal ganglion—"brain."
26. Circumoesphageal connective.
27. Nerve cord.
28. Nerves to body wall and internal organs.
29. Testes.
30. Seminal vesicles—sperm sacs.
31. Sperm funnels.
32. Ovary.
33. Ovarian funnel.
34. Clitellum.
35. Muscle capsule of setae.
36. Setae.
37. Septa.

*The first 22 somites or segments of the Hubbard Earthworm Model shown at left are indicated as S-1, S-2, etc.

Earthworms have a variety of sense organs scattered across their body surfaces, with the greatest number at the two ends. They have a keen sense of touch and can detect light and chemicals within the soil. However, they like the dark and usually come to the surface for feeding and other functions only at night.

Although earthworms have tiny brains, they sometimes seem to demonstrate a fair amount of intelligence in the way they gather their food. Each worm is equipped with both male and female sexual organs, but most species require a pair for reproduction.

Eggs are deposited in a cocoon that incubates in the soil for several months. New worms generally will only emerge when soil moisture and temperature conditions are right.

Garbage Eaters

All earthworms eat organic material in various stages of decay and incorporation into the soil. Normally, they don't feed on living plant tissue, newly planted seeds or seedlings. On occasion, however, they feed on seeds or seedlings that are deteriorating due to emergence problems related to deep seeding, cold, wet soil or soil crusting.

If small, threadlike earthworms are found eating corn seeds or seedlings in a field, they might be mistaken for seed corn

DISAPPEARING CORN. Mike Wurmnest of Deer Creek, Ill., snapped these photos of corn stalks which had been pulled down into worm holes by earthworms. The corn had dried and the worms were eating the sides of the leaves which were pulled 2 to 3-in. into the soil. At right, note the pile of old soybean stubble and corn stubble piled up by worms over nightcrawler mounds.

maggots. However, the seed corn maggot is tapered, smaller at the head and much wider at the rear than the earthworm. Worms are tapered at both ends and have segmented bodies.

During the last week of October one year, Mike Wurmnest of Deer Creek, Ill., was walking through a field of no-till corn that had been planted into bean stubble the previous spring.

"I walked into the field to see if it was dry enough to harvest," he recalls. "The lower leaves of the corn plants were stretched out to nightcrawler mounds. It appeared nightcrawlers had eaten most all of the old bean stubble (only a few stems were left) and were still hungry."

The lower corn leaves had dried and the worms had grabbed

the tips, dragged them down their holes and were eating the sides of the leaves.

"The mid-rib was extending into the soil 2 to 3-in. with the sides of the leaves eaten off," Wurmnest says. "After seeing that, I ran home immediately and told my Dad: We need to get over there and combine that corn before the earthworms pull it down!"

Three Broad Categories

There are 1,800 species throughout the world, but the most common worms in Europe, western Asia and North America belong to the Lumbricidae family. Glossoscelids predominate in Central and South America, while megascolecids are the most common throughout eastern Asia and Australia.

The different species occupy different ecological niches and can be grouped according to their feeding and burrowing habits. Three broad categories have been identified:

➡ *Litter Dwellers.* These small, dark worms live in plant residues or manure piles. They burrow into the soil only when subjected to a stressful environment (such as drought or extreme cold). They have very little impact on soil and crops.

The manure or redworm, *Eisenia foetida,* is an example of an earthworm adapted to live and feed in rapidly decomposing organic matter such as manure or decaying crop residue. It require muck soil with a high carbon content or soil that's been treated with manure to survive.

➡ *Topsoil Dwellers.* These medium-sized worms feed on organic matter dispersed among the mineral particles of the soil. They dig horizontal tunnels in the first 2-in. of soil.

➡ *Subsoil Dwellers.* The largest of earthworms, these create deep vertical tunnels, which allow them to come to the surface to feed on crop residues and deposit their wastes. They may pull leaves down into their burrows to feed on later. The nightcrawler, *Lumbricis terrestris,* is among this group.

Like most nightcrawlers, the most typical garden worms *(Aporrectodea trapezoides, A. tuberculata, A. turgida)* are often used as fishbait. They occupy the top 12-in of soil and move vertically and horizontally.

See the next chapter for more information and drawings of earthworm species commonly found in North America.

Organic, Mineral Matter

Organic and inorganic material (such as fertilizer, pesticides, lime and phosphorus) are selectively ingested by the earthworm and mixed in its gut. Because it is something like a biological "blender," the earthworm plays an important role in maintaining the dynamic equilibrium of soil.

Soil is the uppermost layer of the earth's crust and it represents an ever-changing environment. The soil surface and shallow subsurface includes plant litter, roots, other underground parts of plants and a variety of living organisms. As you dig deeper, you find fewer

---

---

and fewer living things until finally you reach the zone of lifeless mineral formations.

But there rarely are sharp boundaries between the living and non-living soil environments. Because they move throughout these environments, earthworms, especially the subsoil dwellers, help distribute nutrients throughout the soil.

What the earthworm excretes is called a cast. Casts are deposited at the soil surface and generally contain more organic matter and available nutrients than the surrounding soil. Often casts contain higher concentrations of calcium, magnesium, potassium and phosphorus and the pH of casts is close to neutral.

Casts increase the water-stable aggregation of soils. However, in some pasture situations, earthworm activity may actually increase erosion because material otherwise protected by growing grass is exposed on the soil surface.

Freshly deposited casts are not very stable, but with aging and drying they increase water stability. Casts and mucus stabilize earthworm burrows, creating well-aerated soil zones with higher concentrations of organic matter, available nutrients and beneficial microorganisms.

In fact, there may be a zone of increased microbial activity 2-millimeters thick around burrow walls which may increase root growth and nutrient uptake.

Sources

"A Long, Tall Texas Tale," Yleana Martinez, *American Journalism Review,* May, 1993.

Earthworms And Agricultural Management, Eileen Kladivko and H.J. Timmenga, Westview Press, 1990.

Earthworms, Their Ecology and Relationships With Soils And Land Use, K.E. Lee, Academic Press, 1985.

Earthworm model (page 14), Hubbard Scientific, Inc., 1120 Halblieb Rd., Chippewa Falls, WI 54729.

"Worms Are Cleaning Up," *No-Till Farmer,* Mid-March 1993.

Raising Earthworms, Agriculture Canada, Communications Branch, Ottawa, Canada K1A 0C7.

Sustainable Farming, Univ. of Minnesota Extension Service, Fall/Winter 1993.

Identifying Earthworms

THE EARTHWORM diagrams which are found on the next few pages were prepared especially for this book by John W. Reynolds, a world-renowned earthworm researcher working at Sir Sandford Fleming College in Lindsay, Ont.

These earthworms represent eight species commonly found throughout North America, including *Lumbricus terrestris,* the nightcrawler.

With these drawings and the accompanying taxonomic key and glossary (pages 20 to 39), you

---

"10 large worm holes and 100 small holes per square foot means over 330 miles per acre of pencil thickness holes and over 800 miles per acre of small holes...."
—Jim Kinsella, Lexington, Ill.

---

should be able to identify most of the earthworms which you find in your fields.

Know The Worms

Knowing what worms are present can help you understand exactly what's going on in your soils. Plenty of nightcrawlers mean you're getting the full benefits of earthworm activity as they bring up nutrients from the deepest parts of the ground. Many topsoil dwelling worms are a positive sign, but remember they don't travel down as far in the soil as nightcrawlers.

Check 'Em Out

Get a good magnifying glass and a ruler and with this section in hand, check out a worm you have picked up...maybe after a good hard rain.

NOTE: For those unfamiliar with the metric system, multiply centimeters by .4 and millimeters by .04 to obtain length in inches.

TAXONOMIC AND KEY CHARACTERS

The following glossary of morphological structures can be used for the determination of most species of earthworms found naturally occurring in the northern United States and Canada. This list is by no means exhaustive, but is sufficient for the beginner. The drawings are modified from Reynolds (1976a, 1976b, 1977) and Reynolds and Eggen (1993a and b).

A, B, C, D

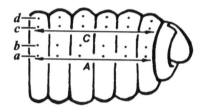

These single capital letters refer to the meridians of longitude passing antero-posteriorly along the apertures of the respective setal follicles. Thus, *A* represents a line along the *a*, the most ventrally located setal follicles.

clitellum

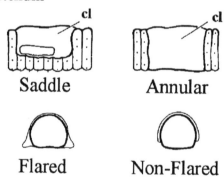

Saddle   Annular

Flared   Non-Flared

A regional epidermal swelling, where gland cells secrete material to form the cocoon. There are two types recognizable. An **annular** clitellum or cingulum encircles the body, whereas a clitellum that encompasses only the dorsal and lateral parts of the body is referred to as a **saddle**. The convention xxvi, xxvii-xxxii, xxxiii means that the clitellum is generally found on the 27th-32nd segments, but in some individuals overlaps onto segments 26 and/or 33. [**cl** = clitellum]

dorsal pore

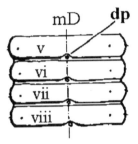

Small single intersegmental apertures in the mid-dorsal line (**mD**) leading to the coelomic cavity. The convention first dorsal pore 5/6 means that the dorsal pore is found in the intersegmental furrow between segments v and vi. [**dp** = dorsal pore]

female pores

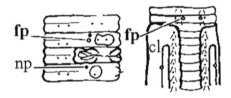

The external openings for the oviducts on segment xiv (Lumbricidae) and ventrad of the mid-lateral line. They are usually more difficult to see than the male pores.

genital tumescences

In Lumbricidae, areas of modified epidermis (glandular swellings), without distinct boundaries and through which follicles of genital setae open. [**cl** = clitellum; **GT** = genital tumescences; **TP** = tubercula pubertatis]

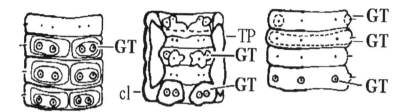

intersegmental furrow

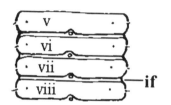

The boundary between two consecutive segments; the area where the epidermis is thinnest and where, in pigmented species, colour is lacking. [**if** = intersegmental furrow]

male pores

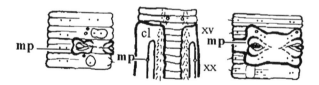

The external openings for the male ducts through which sperm are liberated during copulation. In Lumbricidae, they are usually conspicuous near mL on segment xv. [**mp** = male pore; **cl** = clitellum]

peristomium

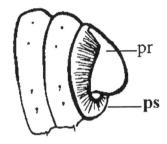

The first body segment, asetal, and containing the mouth.

[**pr** = prostomium; **ps** = peristomium]

prostomium (pr)

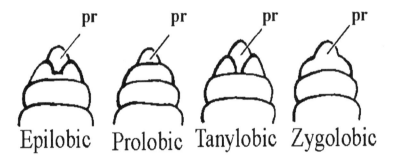

The anterior lobe projecting in front of the peristomium and above the mouth. There are 4 types as seen in the dorsal view above: 1) **Epilobic**: tongue of the prostomium partly divides the peristomium. 2) **Prolobic**: prostomium demarcated from the peristomium without a tongue. 3) **Tanylobic**: with a tongue that completely divides the peristomium. 4) **Zygolobic**: prostomium not demarcated in any way.

secondary annulation

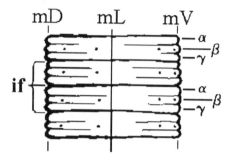

The furrows which occur between the intersegmental furrows (**if**). These demarcations are only external and are labelled α, β, or γ. [mD = mid-dorsal line; mL = mid-lateral line; mV = mid-ventral line]

segment

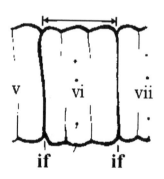

A portion of the body, along the anteroposterior axis, between two consecutive intersegmental furrows (**if**) and the associated septa.

Segments are numbered with lower case Roman numerals, i, ii, iii, *etc.*, beginning anteriorly, with the peristomium as i.

The older system and some microdrile workers still use upper case numerals, I, II, III, *etc.*
[**if** = intersegmental furrow]

seta

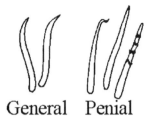

General   Penial

A solid rod or bristle secreted by cells at the ental end of a tubular epidermal ingrowth, the setal follicle. Setae are of several types: 1) **general**: sigmoid shape with pointed outer tip; 2) **genital**: associated with genital tumescences and/or gonopores, and not sigmoid; 3) **penial**: associated with male pores and not sigmoid.

Individual setae are referred to as *a*, *b*, *c*, *d* as shown, *a* being the most ventral and *d* being the most lateral on a particular segment.

setal formula

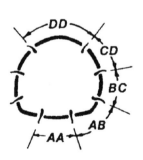

The distance between the setae, usually measured on segments x (10) and/or xxx (30), and being an estimate of the space between the *A*, *B*, *C*, and *D* meridians. The data can be expressed as a ratio (e.g. *AA:AB:BC:CD:DD* = 9:3:6:2:30), as groupings (e.g. *AA>BC<DC, AA = BC*) or in terms of the circumference, C, (e.g. *DD* = ½*C*).

Note: The C for circumference is not in italics, but the meridian lines are in italics.

setal pairings

Setae may be closely paired, widely paired, or separate.

tubercula pubertatis

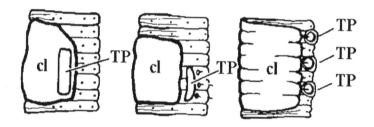

A glandular swelling (**TP**) appearing near the ventrolateral margins of the clitellum (**cl**). It is not always present, and it may be continuous or discontinuous, and of varied size and shape.

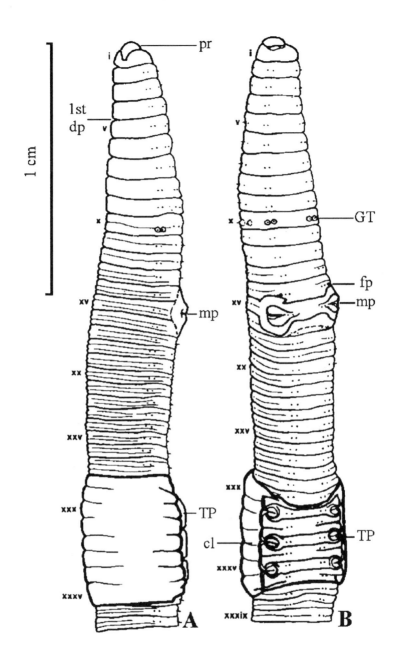

Allolobophora chlorotica (Savigny, 1826) - green worm
A. Dorsolateral view. B. Ventrolateral view

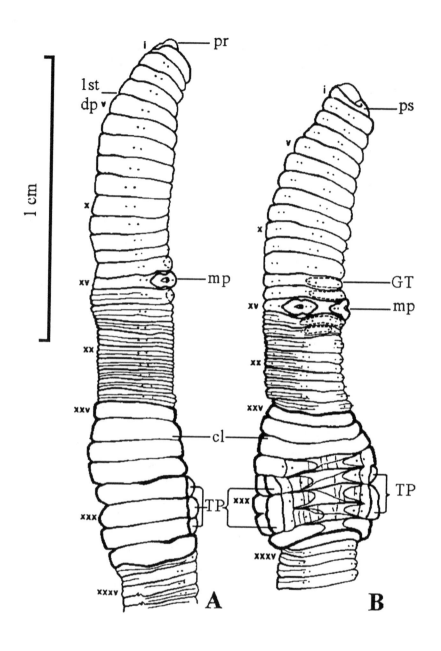

Aporrectodea rosea (Savigny, 1826) - pink soil worm
A. Lateral view. B. Ventrolateral view

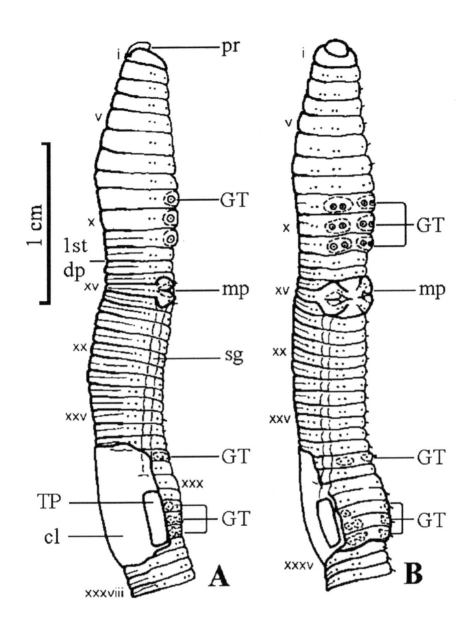

Aporrectodea trapezoides (Dugès, 1828) - southern worm
A. Lateral view. B. Ventrolateral view

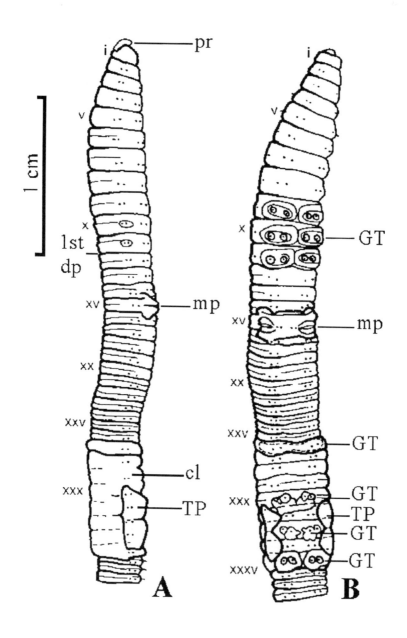

Aporrectodea tuberculata (Eisen, 1874) - Canadian worm
A. Lateral view. B. Ventral view

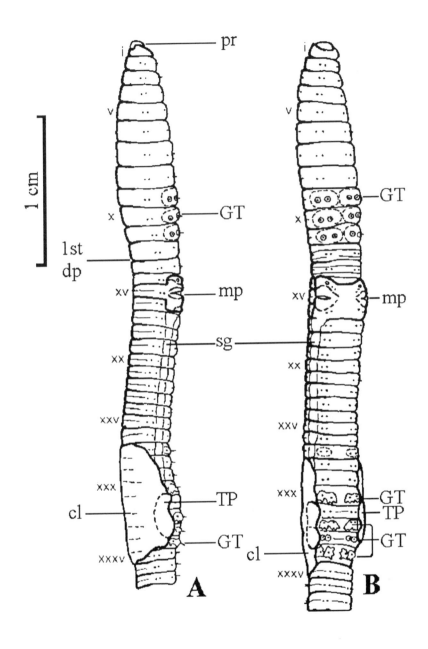

Aporrectodea turgida (Eisen, 1873) - pasture worm
A. Lateral view. B. Ventral view

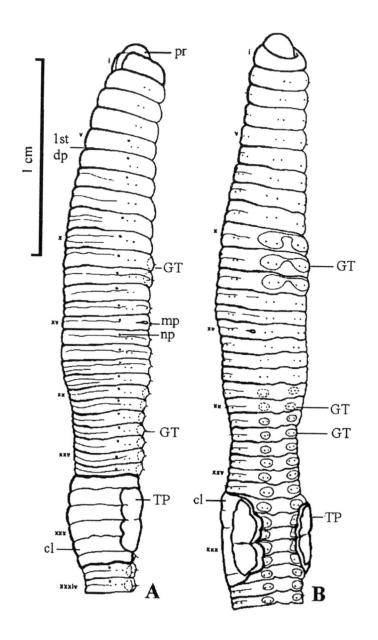

Lumbricus rubellus Hoffmeister, 1843 - red marsh worm
A. Lateral view. B. Ventrolateral view

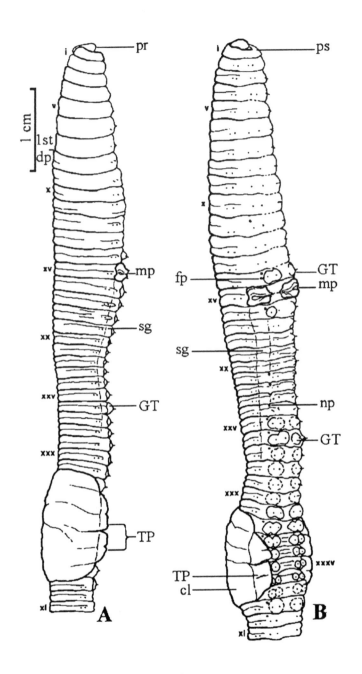

Lumbricus terrestris Linnaeus, 1758 - nightcrawler
A. Lateral view. B. Ventrolateral view

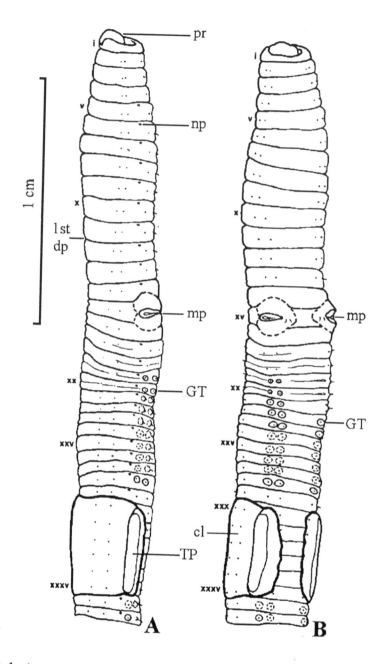

Octolasion tyrtaeum (Savigny, 1826) - woodland white worm
A. Lateral view. B. Ventral view

Species' Diagnoses

Allolobophora chlorotica (Savigny, 1826) green worm

Length: 30-70 mm
Diameter: 3-5 mm
No. segments: 80-140
Colour: generally green, but sometimes yellow, pink or grey
Prostomium: epilobic
First dorsal pore: 4/5
Clitellum: xxviii, xxix-xxxvii (28, 29-37)
Tubercula pubertatis: xxxi, xxxiii, xxxv (31, 33, 35), sucker-like discs
Setae: closely paired, $AA>BC$, $DD = \frac{1}{2}$ C

Aporrectodea rosea (Savigny, 1826) pink soil worm

Length: 25-85 mm
Diameter: 3-5 mm
No. segments: 120-150
Colour: unpigmented, but colour appears rosy or greyish when alive, and white when preserved
Prostomium: epilobic
First dorsal pore: 4/5
Clitellum: xxv, xxvi-xxxii (25, 26-32), somewhat flared ventrally
Tubercula pubertatis: xxix-xxxi (29-31)
Setae: closely paired, $AA>BC<DD$, $AB>CD$, anteriorly $DD = \frac{1}{2}$ C, posteriorly $DD = \frac{1}{3}$ C

Aporrectodea trapezoides (Dugès, 1828) southern worm

Length: 80-140 mm
Diameter: 3-7 mm
No. segments: 93-169
Colour: variable, often lighter behind clitellum until near hind end, then deeper, slate, brown, brownish, reddish brown and occasionally almost reddish, but not purple
Prostomium: epilobic
First dorsal pore: 12/13
Clitellum: xxvii, xxviii-xxxiii, xxxiv (27, 28-33, 34)
Tubercula pubertatis: xxxi-xxxiii (31-33)
Genital tumescences: *a* and *b* setae only, in ix-xi (9-11), xxxi and xxxiii (32-34)
Setae: closely paired, $AA>AB$, $DD< \frac{1}{2}$ C

Aporrectodea tuberculata (Eisen, 1874) Canadian worm

Length: 90-150 mm
Diameter: 4-8 mm
No. segments: 146-194
Colour: unpigmented, almost white or greyish or sometimes with light pigmentation on the dorsum
Prostomium: epilobic
First dorsal pore: 11/12 or 12/13
Clitellum: xxvii-xxxiv (27-34)
Tubercula pubertatis: xxx, xxxi-xxxiii, xxxiv (30, 31-33, 34)
Genital tumescences: absent in xxxi and xxxiii (31 & 33), present in xxx, xxxii and xxxiv (30, 32 & 34)
Setae: closely paired, $AB \approx CD$, $AA > BC$, $DD \approx \frac{1}{2} C$

Aporrectodea turgida (Eisen, 1873) pasture worm

Length: 60-85 mm
Diameter: 3.5-5 mm
No. segments: 130-168
Colour: unpigmented, with the region anterior to the crop flesh pink and the remaining segments pale grey, or occasionally with light pigmentation on the dorsal surface
Prostomium: epilobic
First dorsal pore: 12/13 or 13/14
Clitellum: xxvii, xxviii, xxix-xxxiv, xxxv (27, 28, 29-34, 35)
Tubercula pubertatis: xxix-xxxiii (29-33)
Genital tumescences: *a* and *b* only in xxx, xxxii-xxxiv (30, 32-34) and frequently in xxvii (27)
Setae: closely paired, $AA:AB:BC:CD:DD = 3:1:2:\frac{2}{3}:10$

Lumbricus rubellus Hoffmeister, 1843 red marsh worm

Length: 50-150 mm (usually > 60 mm)
Diameter: 4-6 mm
No. segments: 70-120
Colour: ruddy brown or violet-red, pale yellow on venter (bottom)
Prostomium: tanylobic
First dorsal pore: 5/6 - 8/9
Clitellum: xxvi, xxvii-xxxi, xxxii (26, 27-31, 32)
Tubercula pubertatis: xxviii-xxxi (28-31)
Setae: closely paired, $AA > BC$, $AB > CD$, $DD = \frac{1}{2} C$

Lumbricus terrestris Linneaus, 1758 nightcrawler, dew worm

Length: 90-300 mm
Diameter: 6-10 mm
No. segments: 120-160
Colour: heavily pigmented, brownish-red, or violet on dorsum (top) and yellowish-orange venter (bottom)
Prostomium: tanylobic
First dorsal pore: 7/8
Clitellum: xxxi, xxxii-xxxvii (31, 32-37)
Tubercula pubertatis: xxxiii-xxxvi (33-36)
Setae: enlarged and widely paired at the ends and closely paired in the middle region, $AA>BC$, $AB>CD$, $DD=$ ½ C anteriorly, $DD<$ ½ C posteriorly

Octolasion tyrtaeum (Savigny, 1826) woodland white worm

Length: 25-130 mm
Diameter: 3-6 mm
No. segments: 75-150
Colour: variable, milky white, grey, blue, or pink
Prostomium: epilobic
First dorsal pore: 9/10-13/14
Clitellum: xxx-xxxv (30-35)
Tubercula pubertatis: xxxi-xxxiv (31-34)
Setae: closely paired, anteriorly $CD<AB<BC<AA<DD$, posteriorly $AB>BC>CD$

KEY

The following key is for sexually mature earthworms. Determination of immature forms requires dissection and considerable knowledge of earthworm anatomy and methods of reproduction.

1. Tuberculata pubertatis small sucker-like discs, usually on segments 31, 33 and 35; clitellum on segments 28, 29-37 .... *Allolobophora chlorotica*

    Tuberculata pubertatis not as above 2

2. Prostomium tanylobic 3

    Prostomium epilobic 4

3. Clitellum on segments 26, 27-31, 32; tuberculata pubertatis on segments 28-31 .... *Lumbricus rubellus*

    Clitellum on segments 31, 32-37; Tuberculata pubertatis on segments 33-36 .... *Lumbricus terrestris*

4. Clitellum on segments 30-35 .... *Octolasion trytaeum*

    Clitellum begins before segment 30 5

5. Clitellar region often flared ventrally (bell-shaped in cross section); clitellum on segments 25, 26-33; tuberculata pubertatis on segments 29-31; first dorsal pore in 4/5 .... *Aporrectodea rosea*

    Clitellar region not flared; 1st dorsal pore behind 5/6 6

6. Tuberculata not present before segment 30; clitellum on segments 27, 28-33, 34; genital tumescences often present on segment 28 .... *Aporrectodea trapezoides*

    Tuberculata pubertatis present before segment 30 7

7. Clitellum on segments 27-34; tuberculata pubertatis on segments 31-33; genital tumescences often present on segment 26 and usually absent on segment 33 . . . .
. . . . *Aporrectodea tuberculata*

Clitellum on segments 27, 28, 29-34, 35; tuberculata pubertatis on segments 31-33; genital tumescences often present on segments 27 and 33 . . . . *Aporrectodea turgida*

LITERATURE CITED

Reynolds, J.W. 1976a. The distribution and ecology of the earthworms of Nova Scotia. Megadrilogica 2(8): 1-7.

Reynolds, J.W. 1976b. Catalogue et clé d'identification des lombricidés du Québec. Nat. Can. 103(1): 21-27.

Reynolds, J.W. 1977. The earthworms (Lumbricidae and Sparganophilidae) of Ontario. Life Sci. Misc. Publ., Roy. Ont. Mus., *ix* + 141 pp.

Reynolds, J.W. and A.B. Eggen. 1993a. Earthworm Biology and Vermi-composting I. Lindsay: Sir Sandford Fleming College, 72 pp.

Reynolds, J.W. and A.B. Eggen. 1993b. Earthworm Biology and Vermi-composting II. Lindsay: Sir Sandford Fleming College, 78 pp.

Quotable Quotes...

"We may be able to replace some of the mechanical energy inputs for conventional tillage with the increased activity of soil biological populations..."

—Eileen Kladivko,
Purdue Univ.
West Lafayette, Ind.

"You can't let any one thing, like simply having many more earthworms, actually control you..."

—Dick Thompson,
Boone, Iowa

"Turning in any kind of cover or perennial sod crop will have a major benefit to earthworms..."

—Patrick Bohlen,
Institute for Ecosystem Studies
Millbrook, N.Y.

"Loss of surface litter and a general decline in organic matter content leads to a reduction in the food supply for earthworms..."

—Ken E. Lee,
Glen Osmond, South Australia

"If you're tilling when the worms are emerging from their cocoons, they don't have the body weight to survive while they hunt for food and build their burrows..."

—Dennis Linden,
Univ. of Minnesota
St. Paul, Minn.

"Let's talk about what's really important. I'm taking care of compaction and maintaining soil life."

—Dean Holst,
LeClaire, Iowa

"My no-till fields just love earthworms..."

—Gary Pennell,
Waterville, Ohio

"An increase in earthworm numbers really shows up in the fourth or fifth year of no-tilling..."

—Allen Dean,
Bryan, Ohio

"Even after earthworms die, their holes are still working for you..."

—Jim Kinsella,
Lexington, Ill.

"Earthworms along with terraces can really make a big difference in drainage..."

—Nick Moore,
Shabbona, Ill.

What Earthworms Need To Live

EARTHWORMS FEED on dead, decaying tissue and soilborne micro-organisms. Dead plants are their primary source of food and vary greatly in physical and chemical composition, palatability and nutritional quality.

Because earthworms cannot move very fast, they can't range over a wide territory in their search for food. They must live close to food sources.

"Give them a smorgasbord of crop residues to choose from, with different types of leaves and root structures," says Don Elston, a specialty bean grower from Grover Hill, Ohio. "Rotations with legumes give them wonderful organic matter to thrive on."

Jim Kinsella, a Lexington, Ill., no-tiller, has found earthworms seem to prefer soybean residue to corn residue. This supports the idea that earthworms will thrive better in a corn-soybean rotation than in continuous corn.

Adequate Moisture

Next to a nearby, abundant food supply, earthworms need adequate moisture. They do not have an advanced system of conserving body moisture—in fact, they lose a lot of water in their urine. Since earthworms can only absorb moisture through their

> *"What's important is a good environment in which earthworms survive..."*
> —Yoshikazu Kawaguchi, Osaka, Japan

body walls, those walls must be kept moist.

"Because earthworms need a lot of moisture and cool temperatures, their usual peak activity is in early spring and mid-fall," says Ben Stinner, Ohio State Univ. soil ecologist. "When it's dry during

EARTHWORM HEAVEN. *An ideal earthworm habitat is found in many deciduous forests. Quite frequently, the top layer of soil found in this type of environment is composed almost entirely of earthworm casts.*

the summer, they become inactive to moisture conservation."

LAWS OF EARTHWORM SURVIVAL

1. Give them three square meals a day.
2. Put their food on top of the table.
3. Keep a roof over their heads.
4. Leave them alone.

—*Dick Thompson, Boone, Iowa*

However, if the soil is not well drained, too much moisture will suffocate worms. Earthworms cannot survive in very low or no oxygen environments, although different species have different oxygen requirements.

Comfort Zones

Different earthworm species have different temperature limits for survival. In general, earthworms can survive long exposure to temperatures between 0 to -35 C. If the soil is too warm, earthworms experience stress and their bodies may dry out.

The Air They Breathe

Earthworms absorb oxygen

EARTHWORM BURROWS
AT VARIOUS SOIL DEPTHS

Site Data	Soil Depth (cm)	Burrows Per Square Meter
Orchard, Netherlands	25	200
Pasture, France	Surface	800
Orchard, Netherlands	10	27
	20	22
	30	17
	40	16
	50	33
	60	18
	70	7
	80	1
Cropland, U.S.	7.5	117*
Corn Fields, Germany	2	117*
	20	141*
	30	305*
	60	363*

Land was no-tilled for 4 years.

—*K.E. Lee*

from the air or from oxygenated water. They are threatened by prolonged exposure to dry air because respiration involves loss of body moisture unless humidity is close to 100%. In order to absorb oxygen from water, there must be sufficient concentration of dissolved oxygen.

Sensitive To Light

Light can kill earthworms, especially ultraviolet light. More heavily pigmented species can tolerate more light, but all earthworms will retreat to darkness if given the chance.

Soil Texture

You don't find many earthworms in coarse-textured soils. That's probably because coarse soil particles scrape their tender bodies and also coarse soils tend to dry out faster.

Few earthworms are found in

REACHING A HAPPY MEDIUM. Earthworms often have difficulty surviving in either coarse-textured or wet, high clay soils. But most soil types, like the silt loam profile shown here, fall between the extremes.

high clay soils in high rainfall areas because these tight soils can be seasonally or permanently low in oxygen.

Earthworms, Electrolytes

If soil has a pH of less than 4, you won't find many earthworms. But soils can range from slightly acidic to slightly alkaline without worms being affected much. Chloride and calcium concentrations are often more important to earthworm survival than pH.

Redox potential, the measure of the balance between oxidization and reduction in the soil, also has an impact on earthworm populations. Some ionic and non-ionic substances are toxic to earthworms and can only be toler-

HOW EARTHWORM POPULATIONS VARY

(Silty clay loam soils near Lafayette, Indiana)

Crop Or Condition	Management Practice	Earthworms Per Meter
Continuous corn	Plow	10
Continuous corn	No-till	20
Continuous soybeans	Plow	60
Continuous soybeans	No-till	140
Bluegrass-clover	Alleyway	400
Dairy pasture	Manure	340
Dairy pasture	Heavy manure	1,300

—Purdue Univ.

ated at low concentrations.

Classic Earthworm Soil

The classic habitat of lumbricid earthworms is found in European deciduous forests. It's a type of humus called mull and can be described as a surface layer of soil composed of plant residue mixed with mineral material. Its upper portion consists almost entirely of earthworm casts.

Mull has a low carbon to nitrogen ratio, organic matter with a low proportion of cellular components, pH slightly acid to neutral, high bacterial counts, no matted fungus mycelia, small arthropod populations and many earthworms of various species.

At the opposite end of the spectrum is mor, peaty topsoil with a high carbon to nitrogen ratio, poorly decomposed organic mat-

ter, low pH, low bacterial counts, surface layers matted with fungus mycelia, high arthropod populations and maybe only one or two earthworm species living in small numbers near the surface.

Most humus falls somewhere between these two types. Some earthworms appear in most fields, but may be hard to find in some soils, says former Ohio State Univ. entomologist John Baker.

Sources

Biology Of Earthworms, C.A. Edwards and J.R. Lofty, Chapman and Hall, 1977.

"Earthworms: Digging Out The Facts," Darrell Smith, *Farm Journal*, Mid-January 1990.

"Earthworms First!" Greg Bowman, *The New Farm*, January 1993.

Earthworms, Their Ecology And

Relationships With Soils And Land Use, K.E. Lee, Academic Press, 1985.

"Effect Of Diet On The Feeding And Casting Activity Of Lumbricus Terrestris And Lumbricus Rubellus In Laboratory Culture," M.J. Shipitalo, R. Protz and A.D. Tomlin, *Soil Biology And Biochemistry,* Vol. 20, No. 2, 1988.

Raising Earthworms, Agriculture Canada, Communications Branch, Ottawa, Ont. K1A 0C7.

HOW DIETS IMPACT EARTHWORM WEIGHT, FOOD CONSUMPTION AND CAST PRODUCTION

Diet	Weight Change *% Initial Weight*	Food Consumption *Per Day*	Cast Production *Total Number*
Lumbricus terrestris...			
Alfalfa leaves	35%	13 mg. g*	119
Red clover leaves	19%	12	79
Corn leaves	3%	6	81
Bromegrass leaves	-11%	2	55
No food	-11%	—	60
Lumbricus rubellus...			
Alfalfa leaves	100%	52 mg. g	57
Red clover leaves	76%	36	55
Corn leaves	6%	18	25
Bromegrass leaves	-41%	5	13
No food	-35%	—	43

* *mg. g equals milligrams of dry food per gram of live worm weight per day.*

—Shipitalo, Protz and Tomlin

Getting More Earthworms In Your Fields

IT'S POSSIBLE you may have a field where there are almost no earthworms. They need a surface food supply and mulch protection, says Eileen Kladivko, Purdue Univ. soil scientist. If crop residues are removed too quickly in the spring or fall, worms may not have the opportunity to eat and breed.

Taking A Worm Count

One common way of getting an idea of how many earthworms you have in your fields is averaging the number of holes you find per square foot in several representative locations. You can also spade up some soil samples and count how many worms are present.

You should always check worm numbers near the center of fields, away from fencerows, waterways or other places where perennial plants grow.

If your populations are low, the good news is shallow dwelling worm populations normally rebound quickly with a little encouragement. You can either adopt a tillage program which leaves a residue cover, like no-till or ridge till, or you can build organic matter content by applying manure or sewage sludge, planting cover crops or rotating cropland into hay or setaside acreage.

With any of these methods, you can expect to see dramatic increase in shallow dwelling worm numbers within 1 or 2 years.

Slow Crawlers

It's another story with nightcrawlers. You may have established a suitable environment for them, but normally

WAYS TO INCREASE EARTHWORM POPULATIONS

1. Leave surface mulch.
   No-till.
   Ridge till.
   Cover crops.

2. Add or grow organic matter.
   Manure.
   Hay.
   Setaside acres.
   Cover crops.

   —*Purdue Univ.*

nightcrawlers move into unoccupied territory by slow overland movement and there may not be any crawlers in adjacent fields, fence rows and roadsides.

They're also very slow breeders, Kladivko says, so if you allow nature to take its course, you may wait 7 to 8 years for crawler numbers to build. That is, if they establish themselves at all.

Fishing For Worms

For these reasons, you may want to consider "seeding" nightcrawlers in fields where there aren't many to begin with.

"A small scale, low cost trial is highly advisable, since we really don't know whether or not nightcrawlers will survive in all soils," she says. Some soils may not support crawlers, regardless of management, because of inherent texture and drainage properties. Extremely coarse sands and high water table heavy clay soils are two examples.

Buying Worms?

You could buy nightcrawlers from wholesale suppliers of fishing worms as Robert Van Hoveln of Milford, Ill., did. He received a government grant to experiment with "seeding" 40 acres with earthworms in 1993. So he bought fishing worms by the thousands from a wholesaler.

Ed Slayton of Lafayette, Ind., bought 10,000 nightcrawlers from a local bait shop in boxes of 500 for 5 cents a worm. His cost was about $5 an acre.

"I've spent more on fertilizer and the worms will do more good," Slayton says.

Jim Kinsella, who farms at Lexington, Ill., got 100,000 mature, ready to breed worms out of Canada for 4 cents each. There's quite a thriving fishbait industry in Ontario, which in 1980 exported 500 million nightcrawlers to the U.S. Earthworm hunters there earn as much as $300 a night collecting worms from pastures and golf courses.

But you can go worm hunting on your own. Kladivko advises collecting crawlers from country roads or pastures on rainy spring nights or mornings.

These are the most likely times

WHERE PERENNIALS ARE. *Windbreaks, fence rows and other areas where perennial plants grow are very good places to find earthworms.*

to find worms feeding or mating above ground. The Canadian professionals use miner's lamps to search for worms in the dark with an open tin can strapped to each ankle.

One can contains hardwood sawdust which, when applied to the hands, helps them get a grip on the slippery creatures. The other, larger can serves as a receptacle for approximately 500 worms. On a good night, a Canadian picker can gather 10,000 worms in 7 hours.

Going Fishing?

Of course, it's common courtesy to ask permission before going "fishing" for worms on someone else's land.

According to a report from Georgetown, Ont., a knockdown, drag-out brawl developed when two groups of worm pickers arrived on the same spot on the same farm at the same time. An argument broke out over who had the right to be there, even though neither group had checked with the farm's owner.

Push came to shove and eventually the police had to be called in. Six ambulances were needed to transport 11 people to the hospital.

Dropping Them In

So how do you seed nightcrawlers? Van Hoveln took a

***ADD MORE ORGANIC MATTER.** Anything that you can do to increase the amount of soil organic matter, like planting cover crops or rotating land into set-aside or pasture, encourages earthworm activity.*

cordless drill equipped with an earth auger, bored a hole every 30-ft. and dropped in worms by the handful.

Maybe you don't want to go to that much effort. If not, Kladivko suggests placing 4 or 5 worms under some kind of mulch every 30 or 40-ft. in the field, preferably on a cloudy, wet, cool day.

Record the location of all the seeded spots and watch them for evidence of midden activity throughout the year.

Automated Seeding

Bill Kreitzer of Advanced Biotechnology, Inc., at Gibson City, Ill., is experimenting with encapsulation of earthworm cocoons so they can be "seeded" with a planter. It takes 2 to 3 years for worms hatched from cocoons to reach reproductive age.

Sources

"As The World Turns," *Farm Show, January , 1994.*

Earthworms And Crop Management, Eileen Kladivko, Purdue Univ. Cooperative Extension Service, 1993.

"Effectively Managing Those Highly Valuable Earthworms When No-Tilling," Eileen Kladivko, National No-Tillage Conference, St. Louis, Mo., Jan. 11, 1994 (available on audio cassette).

"Effects Of Tillage On Soil Biology," *Strategies, Techniques And Tactics Guaranteed To Increase Your No-Till Profits,* Lessiter Publications, 1993.

"The Earthworm Bait Market In North America," A.D. Tomlin, *Earthworm Ecology,* J.E. Satchell (ed.), Chapman & Hall, London, 1983.

"Seeding Earthworms With A Cordless Drill," *Ridge Till Hotline,* March, 1994.

Earthworms And Sustainable Agriculture

ONCE PEOPLE seemed to think of LISA simply as a woman's name. Then it became the acronym for "low input/sustainable agriculture," a new catchword in an era of renewed environmental concerns and rising production costs.

Virtually everyone agrees agriculture should be sustainable. That means protecting the environment and conserving natural resources while remaining economically viable and providing for basic human food and fiber needs. Farmers also have an economic interest in reducing their reliance on purchased inputs, too.

How Low To Go?

According to USDA figures, purchased inputs, including pesticides and fertilizer, now represent as much as 80% of total production costs. The question is, when it comes to lowering inputs, how low can you go and still maintain profitable yields?

For some growers, a commitment to sustainable agriculture means total replacement of chemical weed control and commercial fertilizer by manure, crop rotations, tillage, integrated crop-livestock production and traditional soil conservation practices.

"In natural farming, humans are seen to be part of nature," says Yoshikazu Kawaguchi, a farmer from Osaka, Japan, who avoids fertilizers and pesticides as well as tillage. "The yield is smaller with natural farming, but the food is real, it has more life."

For other farmers, sustainable agriculture means more responsible and efficient use of purchased inputs for continued profitability.

Sharon Thompson and her husband Dick's Boone, Iowa, operation "is not organic agriculture, but it doesn't exclude organic agriculture." However, they have used a postemergence herbicide only once since 1967.

"A farming system with 300 acres can be as profitable as one with 3,000 acres," says Dick Thompson. "And the 300-acre diversified crop and livestock system avoids sending 10 farmers to town."

Different Perspectives

The same difference in perspective applies to earthworms and their role in a sustainable agriculture. Complete elimination of potentially toxic substances certainly makes the soil safe for

---

"I've spent more on fertilizer and the worms will do more good..."
—Ed Slayton, Lafayette, Ind.

---

earthworms. On the other hand, earthworms are resilient organisms and many farmers believe reduced amounts of pesticides and fertilizer can be used without adverse effects on them.

"I prefer to apply anhydrous ammonia on in the spring, rather than the fall," says Art Stieren of West Point, Neb. "We have the time to do this with ridge tillage. When we clean off the anhydrous rig, it will be loaded with nightcrawlers and worms."

Anhydrous ammonia only kills earthworms in the small band where it is injected, affecting only a small percentage of the total population, says Eileen Kladivko of Purdue Univ. In addition, nitrogen fertilizer promotes plant growth which, in turn, means higher earthworm populations.

In general, the impact of agricultural chemicals on earthworm populations varies with the chemical. "Most herbicides used in crop production in the Midwest are harmless or only slightly toxic to worms and should not be a great concern," Kladivko says.

Carbamate insecticides and nematicides are highly toxic to worms, but their effects also can be reduced by keeping the application band as narrow as possible. Organophosphate and pyrethroid insecticides are harmless to moderately toxic to earthworms.

Tillage Takes Precedence

The facts show your choice of tillage system will have a greater impact on earthworm populations than whether you use agricultural chemicals or not. Excessive tillage will eliminate more worms than judicious use of pesticides or fertilizer.

"Some farmers are trying organic farming, but they still cultivate to kill weeds and work the soil to make a seedbed," says Ray Cedoz, a Maumee, Ohio, grower.

"Although they believe in the value of earthworms, they travel over the field approximately 13 times to plant, weed and harvest."

Organically grown crops bring a premium price, but the market is small. Cedoz says that too much tillage represents too much cost in terms of soil disturbance for a limited market.

Dick Thompson also believes that an organic farmer who tills in the fall may do more harm to the soil than a farmer who uses herbicides, but doesn't do fall tillage.

It's matter of trade-offs, no matter what management decision you make. Productive pastures usually have higher earthworm populations than any kind of row cropped field, because organic matter is continually being added to the soil normally without any soil disturbance at all, Kladivko points out. Also, legumes and manure make good worm food.

No-till fields usually will have many times higher earthworm populations than plowed fields because crop residue provides more food and mulch protection for earthworms.

However, there is a broad spectrum of tillage practices between traditional moldboard plowing and pure no-till. "Middle of the road" tillage systems will have earthworm populations intermediate between very high (no-till) and very low (plowed fields).

"The amount and type of sur-

SOIL PITS TELL STORY. Andy Gallagher, a Univ. of Wisconsin earthworm researcher, finds earthworms in a field can encourage soil mixing, improve soil structure, help soil aeration, increase infiltration and help incorporate more crop residues.

face residue cover is the key factor to consider when assessing different tillage practices for establishing conditions which encourage earthworm populations," Kladivko says.

Point Of Agreement

Midwesterner Thompson and Far Eastern farmer Kawaguchi agree on one point: earthworms have to be viewed as one component of a total farming system.

"What's important is a good environment in which earthworms survive," Kawaguchi says. Just releasing extra earthworms or other beneficial organisms into the field won't work.

"You can't let any one thing, like more earthworms, control

SOIL TYPE VS. EARTHWORM POPULATIONS

Scotland's Soils

Type Of Soil	Worm Population*	Worm Species
Light sand	232,200	10
Gravel loam	146,800	9
Light loam	256,800	8
Medium loam	226,100	9
Clay	163,800	9
Alluvium	179,800	9
Peaty acid soil	56,600	6
Shallow acid peat	24,600	5

* Thousands of worms per acre.

—Edwards and Lofty

you," Thompson says. "It's not just economics, earthworms or any one kind of tillage that makes a system work for you."

Sources

"Which Is Better? Alternative vs. Conventional Farming," *Agricultural Research*, October, 1989.

Biology Of Earthworms, C.A. Edwards and J.R. Lofty, Chapman and hall, 1977.

Earthworms And Crop Management, Eileen Kladivko, Purdue Univ. Cooperative Extension Service, 1993.

"Effectively Managing Those Highly Valuable Earthworms When No-Tilling," Eileen Kladivko, National No-Tillage Conference, St. Louis, Mo., Jan. 11, 1994 (available on audio cassette).

"Avoid Obsession With Earthworms," *Ridge Till Hotline*, April, 1994.

"The Satori Approach To Farming," David Suzuki, *Toronto Star*, April 16, 1994.

"Smaller, Profitable Farms Mean Kids And Schools," *Sustainable Agriculture*, Univ. of Minnesota Extension Service, March, 1994.

"Soil Biota As Components Of Sustainable Agroecosystems," P.F. Hendrix, D.A. Crossley, Jr., J.M. Blair and D.C. Coleman, *Sustainable Agricultural Systems*, Soil and Water Conservation Society, Ankeny, Iowa, 1990.

"The Importance Of The Fauna In Agricultural Soils: Research Approaches And Perspectives," D.A. Crossley, Jr., David C. Coleman and Paul F. Hendrix, *Agriculture, Ecosystems And Environment*, Elsevier Science Publishers B.B., Amsterdam, The Netherlands, 1989.

Earthworms And Conventional Tillage

IF THERE EVER was an enduring symbol of agriculture, it's the moldboard plow. It's ingrained in our language—even our common religious heritage, as the Bible speaks of "beating swords into plowshares" as a metaphor for peace.

For centuries, farmers prepared a seedbed by turning over the topsoil, either with some kind of plow or by hand, hoe or stick.

Revolution Underway

But there's been a revolution over the last 30 years. Many growers now see extensive benefits from leaving the soil as undisturbed as possible and planting into stubble left from a previous crop. In 1995, 109 million acres or nearly 38% of total crop acres, were planted under conservation tillage, defined as any tillage system that leaves at least 30% of the residue for last year's crop on the soil surface at planting. Among the many reasons farmers and others tout reduced tillage is an increase in earthworm numbers and the ensuing advantages of improved infiltration and fertility.

The Right To Be A Skeptic

But what if you're not ready to join the apparent stampede toward plow-less farming? Even proponents say the success of no-till, minimum till and ridge till depends on relatively new technology, both in the form of equipment and herbicides.

Maybe you're skeptical of claims that some new whiz-bang new spray can provide adequate weed control without tillage. Maybe you're not at the point where you can afford a fancy new planter or you don't want to jury rig the old one with a lot of attachments (which may or may not work the way they're supposed to). It's a free country—you have the right to be a skeptic.

PLENTY OF EARTHWORMS. Even moldboard plowing can turn up plenty of earthworms as shown by these sea gulls enjoying a quick meal in a freshly worked field. Despite such a big emphasis being placed on conservation tillage, you can still manage your earthworm populations with primary tillage to get the biggest possible return.

Perhaps you feel no-tillage is okay or, under the terms of the current Farm Bill, a necessity on steep, highly erodible land. But can you still reap the benefits of improved earthworm populations in conventionally tilled fields?

Forget Fall Tillage

The answer is a qualified yes. You don't have to completely abandon primary tillage, but you probably will have to modify your practices.

For one thing, spring tillage may be a lot better for earthworms than fall tillage. Because earthworms are generally most active in the late fall and early spring, it's best to leave material on top of the soil for them to feed on between fall harvesting and the spring planting season.

Winter cover crops can take the place of crop residues in a conventional tillage system—and incorporation of the cover crop is even good for worms. "Turning in any kind of cover or perennial sod crop—rye, vetch, alfalfa or clover—will have a major benefit to earthworms," says Patrick Bohlen, formerly of Ohio State Univ. and now with the Institute of Ecosystem Studies in Millbrook, N.Y. "After a spring incorporation, the benefit extends through that growing season, up through the fall."

"Mechanical damage, due to tillage implements, has often been assumed to be the major cause of (earthworm) mortality

EARTHWORM POPULATIONS
IN CULTIVATED PLOTS

Tillage Treatment	Weight Per Square Meter*	Number Per Square Meter
Control (unplowed)	6.28	4.18
Minimal cultivation (spring plow)	6.35	5.06
Minimal cultivation (twice spring plow)	6.5	4.37
Maximal cultivation (spring plow)	8.13	5.30
Maximal cultivation (twice spring plow)	7.35	5.55
Maximal cultivation (fall plow)	8.66	6.39
Maximal cultivation (twice fall plow)	10.3	6.88

*Weight in grams per square meter.

—Edwards And Lofty

and population decline," says Australian scientist K.E. Lee.

"But the loss of surface litter and general decline in soil organic matter content that results from long continued cultivation and cropping, and leads to a reduction in the food supply for earthworms, may be even more important."

New Zealand researcher Alec Mackay and Purdue Univ. agronomist Eileen Kladivko add, "Even though the total time between harvesting and planting is only 6 to 7 months, the two peaks of activity might allow sufficient residue mixing and breakdown to have some effect on soil aeration and temperature."

Delay Cultivation?

Ben Stinner, an Ohio State Univ. ecologist, suggests delaying early summer cultivation, depending on weed pressure, until you can see a decrease in the amount of surface castings.

"If you're tilling when the worms are emerging from their cocoons, they don't have the body weight to survive while they hunt for food and build their burrows," says Dennis Linden, Univ. of Minnesota soil scientist. "You can till when the worms are in the adult stage and not do much damage."

If you want to minimize earthworm losses, check the size of specimens you find in several shovelfulls of soil. "Common field worms will be 1/2 to 3/4-in. long when they emerge from their cocoons," Linden explains. "If the majority of the worms you find are that small, then waiting to till would likely improve their survival rate."

Mulching Has Merit

The ancient practice of spreading straw and manure over tilled fields also may help preserve earthworm numbers. Research in Ohio and Maryland shows clean cultivation of cropland after late summer harvesting results in the freezing of the top 4-in. or more of the soil in winter. This leads to the death of many nightcrawlers, but mulching with straw prevented freezing so that the worms survived.

USDA scientists at the National Soil Tilth Laboratory in Ames, Iowa, found grass and leaf-mulched plots had twice as many earthworms as plots where the previous season's corn stalks were left on the soil.

Because tillage tends to dry out the soil, that means worms lose the moisture they need so much in their living environment. However, in high rainfall areas, this is not as much of a problem as in drier regions.

Andy Gallagher, a Univ. of Wisconsin graduate student in soil science, points out not all soils are conducive to earthworms, whether no-tilled or not.

"Nightcrawlers are the main species that increases and does most of the work in no-till fields," Gallagher says. "They like medium textured and silt loam, but not sandy soils. So you won't see many nightcrawlers in no-till fields in the Central Sands area of Wisconsin."

Sources

Biology Of Earthworms, C.A. Edwards and J.R. Lofty, Chapman and Hall, 1977.

"Earthworms First!" *The New Farm*, January, 1993.

Earthworms, Their Ecology And Relationships With Soils And Land Use, K.E. Lee, Academic Press, 1985.

Earthworms And Rate Of Breakdown Of Soybean And Maize Residues In Soil, Alec D. Mackay and Eileen Kladivko, 1985.

"Effects Of Tillage On Soil Biology," Eileen Kladivko, Indiana No-Till Conference, Indianapolis, Ind., Feb. 18, 1985.

"Nightcrawlers Are Good For More Than Fishing," Al Morrow, *Wisconsin Agriculturalist*, March, 1994.

Earthworms And Minimum Tillage

NO DOUBT ABOUT IT. Less tillage means more earthworms, especially nightcrawlers. The more residue you leave on the surface of the soil, the better.

But you don't have to jump right into pure no-till to start seeing increased earthworm numbers. There are a variety of tillage systems that leave 15% to 30% residue on top of the soil at planting time.

Cutting Back On Tillage

It's becoming more and more common for farmers to maybe disk once before planting or disk in the fall and cultivate in the spring or prepare a seedbed with a chisel plow. There are other combinations of light tillage that involve more passes over the field than no-till, but not nearly as many as conventional tillage.

Field trials in Hungary compared the effects of deep versus shallow tillage on earthworm populations. The investigations showed 1 year after shallow cultivation with disks, earthworm populations at four sites were 83.8% of what they were before cultivation. At plowed sites, worm populations were only 49.2%.

Plenty Of Pore Space

Princeton, Ill., farmer Dave Larson says he's been able to boost populations of earthworms and other beneficial soil organ-

---

"We're not disturbing the life in the soil at all. The earthworms are still there..."
—Tom Vittetoe, Washington, Iowa

---

isms with minimum tillage and a biologically friendly fertilizer program.

"I used to believe the only way to handle crop residue was to

FARMERS ARE NO LONGER BURYING ALL RESIDUE. *Numerous research studies show the benefits of not burying all residue when it comes to reducing costs, boosting yields and curbing costly erosion.*

bury it—the deeper the better," Larson says. Then he became aware of the long-term ramifications of conventional tillage. "The soil erosion taking place, even on deep, dark prairie soils with 2% and 3% slope, was significant," he explains.

No-Till Wasn't The Answer

"Realizing this, we tried the opposite extreme: no-till. We found it worked well on the lighter soils and in more rolling situations. But no-till practices on the poorly drained, dark soils have not produced results as good as other systems. Somewhere between the two extremes is most desirable."

Stopping tillage operations does not automatically create good soil structure. In fact, Larson says, farmers who go no-till without a program to stimulate soil life often become discouraged because their soil turns very hard.

The action of earthworms, insects and bacteria in the soil increases the humus content and aggregation which, in turn, improves soil structure and creates more space for air and water.

WORM BURROWS
VS. CROP RESIDUE PLACEMENT

Residue	After 7 Days Burrows	After 90 Days Burrows	After Rain Burrows
Surface	235	360	313
Incorporated	*	*	*
None	235	266	78

*Burrows were visible between loosened surface soil aggregates, but were uncountable.

Castings were present in all treatments except for no residue treatment after a rain.

—Zachmann and Linden

"You can increase air level in the root zone mechanically with tools such as the rotary hoe, cultivator, ridging cultivator and some other deep tillage equipment," Larson says.

WORMS ARE EASY TO FIND. In most minimum tilled fields, earthworms are normally easy to spot with a little digging into moist soil.

Plenty of pore space creates a better environment for soil organisms. "Soil temperatures are more stable during the growing season," Larson notes.

Heavy Soil Needs Aeration

Even dedicated no-tillers sometimes find it necessary to work heavy, tight soil every few years to loosen it up and let in some air. After 2 years of straight no-till in continuous corn, Dean Holst of LeClaire, Iowa, went in and deep tilled in the fall with a narrow row shank ripper of his own design.

Much of the land Holst farms north of the Quad Cities area is very steep and characterized by heavy clay soil. His ripper, equipped with 7/8-in. shanks, leaves more than 60% of crop residue on top of the soil while deep tilling from 12 to 20-in.

"The Soil Conservation Service people said it disturbed the soil less than if I had applied anhydrous ammonia," Holst says.

"Let's talk about what's really important. I've eliminated spring tillage and cultivation. I'm taking

IMPROVING YOUR SOIL STRUCTURE. Reducing costly tillage operations does not guarantee an improvement in good soil structure. It still takes good management of your soils to gain all of the benefits.

care of compaction and maintaining soil life."

Overgrown Rotary Hoe

West Union, Ill., no-tiller Mike Strohm uses two aeration implements to let more air and water into his soil without disturbing crop residue.

"Mike's soil is showing improvement in all facets, even in organic matter," says Bill Becker, director of Central Illinois Research Farms at Springfield, Ill., who's doing research and consulting work on Strohm's crop farm. "He's cut his soil loss to zero and now is building soil, thanks to greater infiltration and cover crops. He's got more earthworms, greater aerobic life and better root systems."

Becker describes Strohm's Aer-Way aerator as "an overgrown rotary hoe." It features a weighted steel frame with ground-driven, tined rollers that penetrate soil up to 7-in. deep.

Built Own Aerator

Strohm also built his own deep aerator which he uses on corn ground every 2 or 3 years. It consists of a toolbar with 24-in. shanks that penetrate up to 16-in. deep.

"With a normal subsoiler, you have to go back and rip up the field again the next year," Strohm says.

AERATING THE SOIL. *Described as an over-grown rotary hoe, this rig features a weighted steel frame with a series of ground-driven, tined rollers which penetrate the soil up to 7-inches deep working more air and water into the soil without disturbing much of the crop residue.*

Mechanical aerators are a way of making do until earthworms can take over, Becker says. The ultimate goal is to build up earthworm populations.

Sources

"Biosystems And No-Tilling," David A. Larson, National No-Tillage Conference, Jan. 12, 1994, St. Louis, Mo. (available on audiocassette).

"Deep Ripping After Two Years Of Straight No-Till," *No-Till Farmer,* Mid-April, 1994.

"Earthworm Effects On Corn Residue Breakdown And Infiltration, J.E. Zachmann and D.R. Linden, *Soil Science Society Of America Journal,* Nov./Dec. 1989, Vol. 33, No. 6.

Earthworms, Their Ecology And Relationships With Soils And Land Use, K.E. Lee, Academic Press, 1985.

Low Cost, High Yield Farming, David A. Larson, Princeton, Ill.

"Mechanical Earthworms," Michael Traupman, *Country Folks Of Pennsylvania,* April 11, 1994 (reprinted from *New Farm).*

"No-Till Special, Hold The P & K," *No-Till Farmer,* May 1993.

What Really Happens When You Cut Chemicals? Rodale Press, 1993.

Earthworms And No-Tilling

TILLAGE SYSTEMS that leave 30% or more of residue from the previous crop on the soil surface at planting provide the best environment for all kinds of earthworms.

By dispensing with tillage altogether, you not only provide worms with a food supply and protection from the weather, you also avoid disturbing the soil where they build their burrows.

This is particularly beneficial for nightcrawlers, because of their permanent, vertical burrows. These, in turn, improve a no-till field's capacity for absorbing water by serving as conduits through the soil.

No-Till Farmer Survey

More than 60% of *No-Till Farmer* newsletter readers who answered a postcard survey said earthworms flourished in fields.

"My no-till fields just love earthworms," says Gary Pennell of Waterville, Ohio. "They're appearing where I never had them."

Only 8% said their earthworm populations stayed the same or decreased. The remaining 31% either hadn't paid attention to worms or hadn't been no-tilling long enough to notice a difference in populations.

In Univ. of Minnesota research, there were twice as many earthworm burrows open to the soil surface in a no-till plot compared to a tillage with incorpo-

---

HOW FARMERS THINK NO-TILLING AFFECTS FIELDS

More worms	61%
Fewer worms	1%
Same number	8%
Don't know	30%

—No-Till Farmer

MORE NO-TILL, MORE EARTHWORMS. *Numerous research projects have shown the less you till, the more worms you will find in fields.*

rated residue treatment.

In southeastern Indiana, researchers found three times as many earthworms in silt loam in a corn-soybean rotation under no-till than with chisel plowing.

EARTHWORM POPULATIONS UNDER VARIOUS NO-TILL CROP ROTATIONS

(1990-1991 Crop Rotation, Measured In Number Of Worms Per Square Yard)

Variable Measured	Wheat-Corn	Corn-Soybeans	Soybeans-Wheat
Earthworms	257	346	443
Cocoons	27	71	35
Total	284	417	478

—Breker Farm, Havana, N. D.

TRACKING RUNOFF. Bill Edwards, upper left, a USDA researcher located at the North Appalachian Experimental Watershed in Coshocton, Ohio, has been evaluating the presence and movement of earthworms in soils and runoff of chemicals for a number of years. He finds earthworms are very beneficial when it comes to helping surface water get into the root zone and preventing costly runoff.

At the Rothamsted Experimental Station at Harpenden, England, 8 years of observation showed nightcrawler number increased in no-tilled fields while decreasing in moldboard and chisel plowed fields.

"My earthworm population has virtually exploded since I quit killing them with tillage," says Randy Henke of Sawyer, N.D.

"We have one field that's in its 11th consecutive year of no-till," says Allen Dean of Bryan, Ohio. "An increase in earthworm numbers really shows up in the fourth

PESTICIDES VS. EARTHWORMS. At Coshocton, Ohio, Bill Edwards has injected chemical solutions directly into worm holes and then caught the resulting liquid as it flows out the bottom of the holes in the ground. Analysis of the liquids indicates no major environmental concerns from chemicals moving through fields via earthworm holes.

or fifth year of no-tilling."

"Earthworms really increase, especially after a field has been no-tilled for several years," says Tom Doorley, Waponetka, Ohio.

"Earthworms are an important part of no-till—some would say an essential part," says Jim Kinsella, who's been no-tilling at Lexington, Ill., for over 25 years. "I believe you can't get along without them. Earthworms are nature's tillers. They're cheap and do an excellent job."

Improved Water Filtration

USDA researcher Bill Edwards has grown no-till corn since 1964 on a watershed with an average 9% slope at Coshocton, Ohio. With an average annual rainfall of 39.4-in., runoff has averaged .8-in. on the no-tilled area and 4.9-in. on nearby conventionally tilled corn plots.

This is in large part because earthworm burrows form "macropores" in the soil. Rain penetrates the macropores, draining surface water to the root zone and preventing runoff.

"Earthworms along with terraces make a big difference in drainage," says Nick Moore, a Shabbona, Ill., no-tiller. "I believe a farmer should put as much tile in his hills as on flat ground to make the water go under the

BURROWS OPEN TO SOIL SURFACE AFTER 6 WEEKS OF EARTHWORM ACTIVITY

(As Affected By Tillage Treatment And Earthworm Species in Minnesota)

Tillage Treatment	Earthworm Species	Number Of Burrows*
Tilled**	A. tuberculata	14.5
	L. rubellus	15.0
No-till	A. tuberculata	36.5
	L. rubellus	47.5

* Measured in 0.06 square meter subplots inoculated with 13 earthworms each.
** Tilled with rototiller to a 10-centimeter depth.

—Zachmann, Linden and Clapp

ground and not over it."

At the Dakota Lakes Research Farm near Pierre, S.D., research director Dwayne Beck found earthworms in no-till fields helped save on irrigation costs. Thanks to the enhanced infiltration, water application pressure was reduced from 60 psi to 6 psi. That pressure decrease translated into more than a 50% reduction in irrigation energy costs.

After 4 to 5 years of no-till, high populations of earthworms can burrow through the tightest soils. Topsoil turns into a sponge and stops field runoff, Beck says.

"With no-till, you don't disturb the earthworm holes," says Ken Rutz, LaPorte County, Ind., farmer. "These holes hold the water and channel it deeper into the soil. No-till reduces runoff tremendously."

Barrels Of Worms

In 1991, Iowa State Univ. and USDA researchers buried open-ended barrels filled with earthworms in no-tilled fields on two farms in southeast Iowa. Water infiltration rates proved to be four to five times faster in fields seeded with nightcrawlers than where no worms were introduced.

Reducing Runoff

Half a world away in South Australia, state researcher John Buckerfield supervised a similar

A BIG CUMULATIVE IMPACT. The longer you no-till your fields, the greater the population should be of your highly valuable earthworms.

experiment in 1992 which showed how earthworms reduce runoff. Two cylinders were filled with layers of different soils divided horizontally by bands of sand. Worms were introduced into one cylinder while the other remained barren.

Within 3 months, the top two thirds of soil and sand in the

EARTHWORM POPULATIONS AS INFLUENCED BY AMOUNT OF SURFACE RESIDUE LEFT AFTER NO-TILLING FIELDS

Variable Measured Measured	40% to 45% Residue	80% to 90% Residue
Earthworms	71	106
Cocoons	204	514
Total	275	620

—Langdon Research Center, Langdon, N.D.

worm-seeded cylinder was thoroughly mixed. Crop residue placed on top of the soil was broken down and incorporated.

The second cylinder remained unchanged. Then Buckerfield's team watered the cylinders to simulate rainfall.

After water accumulated at the bottom of the wormless cylinder, Buckerfield drilled a hole in the bottom.

"The water was like a sludge containing all the fine silty soil from the upper levels," he says. "In fact, an underground 'cave' appeared where the soil had come from."

In the other cylinder, water moved down through worm burrows, causing very little soil disturbance.

In other research by Buckerfield and other Australian scientists, no-till fields were found to have four times more earthworms than conventionally tilled fields.

Aerating The Soil

"We have 1,200 miles of earthworm holes in every acre," says Kinsella. "Before we went to no-till, there weren't any worms. We had to plant them back into the soil."

Earthworms help aerate the soil, Kinsella says. "Soil bacteria needs air to do its work of breaking down roots, stalks and other matter," he adds. "It can't work if there's no air and without worms, there are no holes to let air into the soil structure."

Earthworms may live as long as 8 or 9 years. "Even after they die, their holes are still working for you," Kinsella says.

After 8 years of no-till, some of his soil with 6% slope withstood a 100-year rain without runoff, Kinsella says.

Some of his fields have 15 large nightcrawler holes per square foot. "Well-managed no-till farming over 10 years should

1,200 MILES OF WORM HOLES.
Veteran no-tiller Jim Kinsella of Lexington, Ill., says that thousands of earthworms help aerate the soil. He finds worms may live as long as 8 or 9 years in his fields. Even after they die, the remaining worm holes are still working for you.

produce 100 small worm holes per square foot going down 12-in. and at least 10 large earthworm holes per square foot down to 4-

---

NO-TILLED VS. PLOWED SOILS

(1967-1969 Study In England)

Tillage And Year	Number Of Worms*	Weight Of Worms**
No-Till...		
1967	431	70.9
1968	431	108.5
1969	344	36.0
Plowed...		
1967	25	2.0
1968	331	43.7
1969	487	25.2

* *Number of worms per square meter.*
** *Weight of worms in grams per square meter.*

—*Edwards and Lofty*

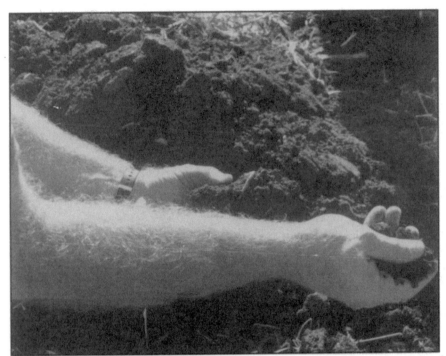

THE SPONGE EFFECT WITH NO-TILLING. *After a number of years of continuous no-tilling, high populations of earthworms can burrow through even the tightest soils, which helps stop costly field runoff.*

ft.," Kinsella says. "This means over 800 miles per acre of small holes and 330 miles per acre of pencil thickness holes."

Sources

"Earthworm Display Turns Out A Winner," Huw Morgan, *The Advertiser,* Sept. 23, 1992.

"Earthworm Effects On Corn Residue Breakdown And Infiltration," J.E. Zachmann and D.R. Linden, *Soil Science Society Of America Journal,* November-December, 1989, Vol. 53, No. 6.

"Earthworms Improve Soil Tilth, Drainage," Vic Virgil, *Warsaw, Ind., Times-Union,* Feb. 2, 1994.

"Earthworms Increase Water Infiltration," *Henderson Co., Ill. Quill,* Jan. 29, 1992.

"Earthworm Populations In Dryland Cropping Soils Under Conservation Tillage In South Australia," John C. Buckerfield, *Soil Biology And Biochemistry,* Vol. 24, No. 12, 1992.

"Earthworm Populations Related To Soil And Fertilizer Management Practices," E.J. Deiberrt and R.A. Utter, *Better Crops,* Vol. 78, No. 3.

"Factors Affecting Preferential Flow Of Water And Atrazine Through Earthworm Burrows Under Continuous No-Till Corn," W.M. Edwards, M.J. Shipitalo, L.B. Owens and W.A. Dick, *Journal Of Environmental Quality,* September, 1993.

Jim Kinsella, Innovative Farmers

TRAM LINES WITH NO-TILL. *Some northern no-till small grain growers are finding permanent tram lines left in their fields are helping reduce compaction, resulting in more earthworms and higher yields.*

Workshop, March 3-4, 1992.

"Macroporous Infiltration And Redistribution As Affected By Earthworms, Tillage and Residue," J.E. Zachmann, D.R. Linden and C.E. Clapp, *Soil Science Society Of America Journal*, November-December, 1987, Vol. 51, No. 6.

Management Of Earthworms In Agriculture, John C. Buckerfield, CSIRO Division of Soils, October, 1993.

"Nightcrawlers Help No-Till," Robert Schiffner, *Monroe, Mich., Evening News*, Jan. 18, 1994.

"No Doubts About No-Till," Tim Stoner, *Valparaiso, Ind., Vidette-Messenger*, March 31, 1992.

"No-Till Methods Play Big Role In Protecting Water Quality," Christine Anderson, *Bloomington, Ill., Pantagraph*, March, 1993.

"No-Till Saves, Builds Soil, Says Advocate," Dawn Johnson, *Rock Island, Ill., Argus*, Sept. 3, 1989.

"Take Care Of Those Nightcrawlers...And They'll Take Care Of Your Soil," *No-Till Farmer*, January, 1993.

"The Effect Of Direct Drilling And Minimal Cultivation On Earthworm Populations," C.A. Edwards and J.R. Lofty, *Journal Of Applied Ecology*, 1982.

"The High Tech Road To Sustainability," Norman Senjem, *The Dakota Farmer*, October, 1990.

"Why Earthworms Are A Big, Big No-Till Bonus," Bill Edwards, National No-Tillage Conference, Indianapolis, Ind., Jan. 13, 1993.

Earthworms And Ridge Tillage

LIKE NO-TILL, ridge till leaves most of the soil profile undisturbed, but has the added benefit of reduced chemical use.

As a row crop production system, ridge tillage involves planting, cultivating and harvesting, but no primary tillage such as plowing, disking or chiseling.

Crops are planted on top of ridges normally built to a height of several inches at second cultivation the previous year. Special ridging cultivators pile soil and crop residue from the furrows between ridges onto the ridgetops.

Most of the residue remains in the valleys, however. Ridges provide a warmer, drier environment for seeds to germinate, while the residue cover provides erosion control benefits and also supplies earthworms with food.

Heading For High Ground

In addition, ridges provide high ground for earthworms during torrential rains. Some ridge tillers report finding more worms on the ridges than in furrows after a downpour.

"In ridge tilled fields, we find earthworms concentrated in the area under the rows and row middles, but not on the sides of ridges," says John Dickey, former

---

"I haven't seen worms in the field since I was a kid tagging along behind Dad's 8N Ford and the two-bottom plow..."
—Kenneth Replogle, Coweta, Okla.

Purdue Univ. agronomist. "This is because the sides of ridges have little cover."

Many ridge tillers put row cleaner attachments on their planters to shave soil and residue from the tops of ridges in the

WHERE ARE THE EARTHWORMS FOUND? *Purdue Univ. research indicates most earthworms tend to concentrate in the area under the rows and row middles rather than being found on the sides of ridges.*

spring. This improves seed to soil contact and pushes freshly deposited weed seeds into the furrows. Sometimes, however, ridge tillers do not use row cleaners and simply "no-till" into the ridges.

Ridge till planters also are often equipped with fertilizer and pesticide application attachments to get as much done in one pass over the field. Other options for ridge tillers include separate fall or spring application of fertilizer and chemicals or application with the cultivator.

Because ridges force farmers to follow the same wheeltracks with every operation, there's the added benefit of reduced seed row soil compaction. Although

soil may become compacted in the wheeltracks, earthworms will still live and work in those areas, notes Dick Thompson, Boone, Iowa, ridge tiller.

Ridge till offers the opportunity to cut both fuel and machinery costs and herbicide expenses with a combination of chemical and mechanical weed control. Yields equal or better than other tillage systems regularly put ridge till operations among the most profitable in production cost surveys like the nationwide Farming for Maximum Efficiency (MAX) program.

In an Ohio State Univ. survey, ridge tillers reported both the highest corn yields and net farm

EARTHWORM POPULATIONS WITH CORN, SOYBEAN ROTATION ON SILT LOAM SOIL

(Measured In April In Southeastern Indiana)

Tillage Practice	Earthworms Per Square Meter		
	1987	1988	1989
Ridge till	—	189	178
No-till	156	133	211
Chisel plow	—	44	67

—Purdue Univ.

incomes despite a season-long drought. No-tillers reported the lowest corn yields and tied with conventional tillers for lowest soybean yields in the same survey.

Kenneth Replogle, one of Oklahoma's first ridge tillers, says his herbicide costs are roughly a third of what he spent before switching to ridge till in 1990.

He's running a tractor less than half the size of his old equipment and using 3 less gallons of diesel fuel to produce an acre of soybeans. Reductions in purchased inputs have paid for the Coweta, Okla., farmer's investment in ridge till equipment.

Big Earthworm Increases

When it comes to earthworm populations, ridge till and no-till run neck and neck. In southeastern Indiana comparisons, earthworm numbers in no-till and ridge till fields were comparable, with ridge till plots showing slightly more worms in 1988 and no-till plots slightly more in 1989.

"Populations will vary from year to year as well as within a year due to weather conditions and food availability," says Eileen Kladivko, Purdue Univ. soil scientist.

Most ridge tillers are happy

CULTIVATION DOES PAY. Farmers who are using less herbicide for ridge tilling find working their soils with a cultivator helps increase earthworm populations.

RIDGE TILL VS. CONVENTIONAL TILLAGE EARTHWORMS

(Per 8-Inch Cylinder)

Ridge till	10.17
Conventional tillage	0.15

—National Soil Tilth Lab

with the increases in earthworm populations that result from ridge till. "I hadn't seen worms in the field since I was a kid tagging along behind Dad's 8N Ford and two-bottom plow," Replogle says.

"I sure put a high value on the extra earthworms we have with ridge till," says Art Stieren of West Point, Neb. Once when Stieren was putting on anhydrous ammonia, the shanks plugged up with residues.

"I went to clean them off and found the shanks loaded with nightcrawlers," he recalls. "It made me wonder whether I should keep on applying anhydrous or go fishing instead."

"Earthworms may be bringing low level lime in fields up to the surface," says Randy Quilliam of Millersport, Ohio. "We are definitely improving both structure and organic matter in our soils."

"We definitely see more earth-worms all the time with our ridge tilling system," says Dean Eppley of Wabash, Ind. "It seems like the reduction in tillage has really helped us increase our earthworm populations. But I'm not sure whether the increase in earth-worms is due more to tillage or to increased fertility."

Thousands Of Worms

After 21 years of ridge till, Ohio State Univ. researchers found an average of 825,000 earthworms per acre on John Alexander's Fountain City, Ind., farm, even though he had used anhydrous ammonia and insecticides. On a neighbor's farm, where continuous corn was grown under conventional tillage for the same period of time, they found hardly any earthworms.

"Whenever you pick up a rock, you will almost always find an earthworm or two under it," Alexander says. When he bought his farm in 1950, corn was yielding

"I went to clean off the anhydrous shanks in my ridge tilled fields and found them loaded with nightcrawlers..."
—Art Stieren, West Point, Neb.

only 50 to 70-bu. per acre. By fertilizing properly, Alexander increased yields to 100-bu. per acre by 1967 when he started ridge tilling. By 1977, corn yields were

WIDE SPACED WHEELS. Running tractor and implement wheels in permanent paths in ridge tilled fields helps reduce compaction concerns, helping boost your earthworm numbers.

averaging 170-bu. over the entire farm.

"It took 4 years of ridge tilling to get water into these soils," Alexander says. "Once the soils were aerated, tilth increased."

Field To Field Comparison

"Sharon and I knew from observation that our fields had less ponding, crusting and clods, more earthworms and were softer when you walked on the soil compared with our neighbors' fields," says Dick Thompson. "But we needed a third party to verify it because we didn't have much hard data to back it up."

Then in 1989, Doug Karlen, USDA soil scientist, led a team of scientists, technicians and graduate students in an effort to com-

RIDGE TILL DOESN'T BOTHER THE EARTHWORMS. Despite using herbicides, insecticides and anhydrous ammonia, many ridge tillers find earthworm populations continue to grow each year in their fields.

DUALS GIVE TRAFFIC PATHS. Anything you can do to curb compaction with ridge till is also likely to help improve earthworm numbers.

pare Thompson's fields with those of his neighbor, conventional tiller Eugene Baker.

After carefully mapping and measuring the fields for soil structure, runoff, infiltration, nitrogen profiles, earthworm populations and other factors, they found Thompson's land was less prone to erosion and crusting. Water infiltration rates tended to be higher and runoff levels lower.

In the fall of 1989, earthworm counts averaged 562 per square meter on Thompson's farm. None were found in Baker's fields.

Fall disking in 1990 in preparation for planting oats reduced counts to 70 worms per square meter in Thompson's fields. That same year, researchers found an average of 9.3 worms per square meter with one soil type in Baker's fields.

Sources

Earthworms And Crop Management, Eileen J. Kladivko, Purdue Univ. Cooperative Extension Service.

"Hard Data On Soil Health," Craig Cramer, *New Farm,* January, 1993.

"If Ridge Till Hadn't Improved His Soils, He Would Have Gone Broke!" *Ridge Till Hotline,* July, 1991.

Making Ridge Till Work For You, William A. Hayes, Lessiter Publications, 1993.

"Manage Your Fields For More Earthworms," Carl Eiche, *Prairie Farmer,* Oct. 3, 1989.

"Ridge Till Coming To Oklahoma," Dan Crummett, *Oklahoma Farmer-Stockman,* April, 1992.

"Ridge Tillers Laugh On Way To The Elevator," *Ridge Till Hotline,* April, 1993.

Thompson On-Farm Research, 1993 Report.

Fine Tuning
Your Fertilization

ANHYDROUS AMMONIA and its effects on soil life remains the biggest concern in fine tuning fertility management to accommodate earthworms. After all, the gas was used to freeze dry and harden soils for building bomber runways in the South Pacific during World War II, says Dave Larson, a Princeton, Ill., farmer.

Minimum Impact

The good news is with a few precautions, you may apply anhydrous with minimum impact on crawlers. Or you can choose from a variety of alternatives to anhy-

---

**SPRING EARTHWORM POPULATIONS
AFTER ANHYDROUS AMMONIA APPLICATION***

Burchill Farm, East Central North Dakota

Measured Variable	0-lbs. N	Sept. 15 50-lbs. N	Nov. 1 50-lbs. N
Earthworms**	98	98	292
Cocoons	62	71	98
Total	160	169	390

* Based on actual nitrogen rates per acre.
** Measured in number of earthworms per square yard.

—Better Crops

NO PROBLEMS WITH ANHYDROUS AMMONIA. *Many conservation tillage farmers say they use anhydrous on a year after year basis and have continued to see earthworm numbers increase in their fields.*

drous application. High concentrations of any ammonia-based fertilizer will kill worms, but improved fertility boosts earthworm populations in the long run. That's because the more fertility, the higher the yield and the higher the yield, the more crop residue for worms to feed on.

"Use any nitrogen source you want to use," says Ernie Behn, ridge till pioneer from Boone, Iowa.

Anhydrous No Problem!

George Rehm, Univ. of Minnesota soil scientist, puts it this way: "If an earthworm gets in front of an anhydrous shank, that sucker's going to die!" But he adds, "We drown more earthworms fishing than we hit with anhydrous ammonia shanks."

Dick Lloyd of Lewiston, Idaho, has used anhydrous ammonia successfully in his small grain operation since it came out in the 1940s.

"I just don't buy the idea that anhydrous ammonia creates a

"I just don't buy the idea that anhydrous ammonia creates a problem with earthworms..."

—*Dick Lloyd, Lewiston, Idaho*

---

problem with earthworms," Lloyd says. In fact, earthworm numbers have increased since he switched to no-till farming because of organic matter buildup and reduced disturbance of the soil.

Worms seem to actually hear

the tractor coming, says Jim Kinsella of Lexington, Ill. "I use NH3 every year and I have no problems with worms. They move deeper into the soil fairly fast when I'm applying it."

Two Rules Of Thumb

There are two rules of thumb for minimizing anhydrous impact on earthworms, says John Dickey, an agronomist who conducted earthworm studies at the Southeast Purdue Ag Center at Butlerville, Ind., as a graduate student. First, try to sidedress where there are no crop residues. Second, apply fertilizer when the top layers of soil are dry.

"When soils are dry, earthworms will be deeper in the soil," Dickey says.

"Ridge tillers will probably want to apply anhydrous or pesticides on the shoulders of their ridges because that's where the earthworms won't be," Dickey says. "However, even when we applied anhydrous in row middles with ridge till, earthworm populations were never reduced more than 5%."

Giving Worms Time Of Day

"I don't sidedress nitrogen in the corn real early in the morning or real late at night," says Greg DeMott, Ashley, Mich. "Those are the times when earthworms are up in the danger zone."

Jon Anderson of Groton, S.D., waits until late fall to apply anhydrous because then the soil surface temperature is around 40 degrees F. and worms are deeply burrowed.

Alternatives To Anhydrous

Some farmers put such a value on earthworms that they avoid anhydrous altogether rather than lose any worms. Jon Larson of Randolph, Kan., stopped using anhydrous 20 years ago for that reason.

"No anhydrous, but use lots of

USE OTHER NITROGEN SOURCES. *Some farmers are concerned enough about potential anhydrous ammonia damage to earthworm populations that they have decided to go with other nitrogen options to protect worms, like surface application of liquid urea solutions.*

manure," suggests Sam Studebaker of Tipp City, Ohio.

"Our nitrogen source on corn is applied over the top so NH3 doesn't damage the earthworm population," says Vince Roberts, Dearborn, Mo. Verle Saxton, Beardsley, Minn., prefers broadcasting nitrogen rather than knifing it in.

Asa Chester of Columbus, Ohio, injects 28% nitrogen rather than applying anhydrous. David Drake, Girard, Ill., notes a 28% solution more readily breaks down residue for the worms.

Bill Holtz, Heyworth, Ill., sidedresses 28% nitrogen. In addition to sidedressing 28% nitrogen, Dan Monroe, Dupont, Ind., spreads chicken manure and municipal sewage to increase worm numbers.

It's been over 15 years since Mike Wurmnest of Deer Creek, Ill., applied anhydrous ammonia. "We strip in 28% nitrogen in the fall or winter, followed by another shot broadcast with herbicides in early spring before it gets really hot," Wurmnest says.

"I have always felt if the soil was loose with a lot of air in it and if it also had a lot of earthworms, then a split application of 28% nitrogen over the top would be easier on the earthworm population," he says.

Research conducted in Egypt, England and Germany suggests that earthworms may be affected more by various changes in soil pH brought about by fertilization than by anything else.

"Nightcrawlers can tolerate pH ranges as low as 4, but 7 or 8 is the optimum pH level," says John Bater, former Ohio State Univ. entomologist now in England.

However, all forms of applied fertilizer generally lead to increased earthworm populations compared to untreated plots.

Most forms of nitrogen pass through the earthworm's gut without much chemical change and thus return to the soil.

Sources

Biology Of Earthworms, C.A. Edwards and J.R. Lofty, Chapman and Hall, 1977.

"Biosystems And No-Tilling," Dave Larson, National No-Tillage Conference, Jan. 12, 1994, St. Louis, Mo.

"Earthworms Increase In Spite Of Anhydrous," *No-Till Farmer,* May, 1993.

"Earthworm Populations Related To Soil And Fertilizer Management Practices," E.J. Deibert and R.A. Utter, *Better Crops,* Summer, 1994, Vol. 78, No. 3.

Earthworms, Their Ecology And Relationships With Soils And Land Use, K.E. Lee, Academic Press, 1985.

"Manage Your Fields For More Earthworms," Carl Eiche, *Prairie Farmer,* Oct. 3, 1989.

"Take Care Of Those Nightcrawlers...And They'll Take Care Of Your Soil," *No-Till Farmer,* January, 1993.

Manure And Livestock Management

EARTHWORMS LOVE manure and it's good for them, too. Some worm species are actually adapted to live and feed in animal wastes. For farmers who raise both crops and livestock, that's a big plus.

All forms of fertilizer contribute to earthworm well-being by increasing yields and therefore crop residue for worms to feed on.

But manure has a greater positive effect on earthworms than commercial fertilizer because it already contains plant residues as well as soil nutrients, says Dwayne Beck, director of the Dakota Lakes Research Farm at Pierre, S.D.

The nutrients earthworms need are present in manure in an already digested form, says Barry Swalla, a researcher at the National Soil Tilth Laboratory in Ames, Iowa. USDA scientists there use a mixture of Iowa soil and horse manure to breed worms for research. Of course, they have to obtain the manure from an owner who doesn't use a wormer on his horses.

Don't Spread It Too Thick

Spread manure too thick and you'll cut off earthworms from their oxygen supply, cautions Andy Gallagher, Univ. of Wisconsin researcher. A southeastern Wisconsin farmer confirms this hazard by saying, "We emptied our lagoons. You never saw so many dead worms."

Dick Thompson, a Boone, Iowa, ridge tiller, says nothing beats applying manure for increased organic matter and larger earthworm populations.

He's devised a unique yield-boosting program of applying hog and cattle manure within 48 hours of ridge tilling his crops.

Subsurface Application

Tom Vittetoe and his brother,

FORAGING FOR FOOD. Concentrated in a barn or yard, spread in fields or dropped by animals in pasture and stalk fields, livestock manure provides an excellent means for boosting the worm numbers.

Jerome, have to do something with a lot of hog manure right outside of Washington, Iowa. Tom raises the offspring of 300 sows farrow-to-finish and Jerome, who's strictly a livestock producer, has 1,000 head in his farrow-to-finish operation. Together, they have to dispose of over 5 million gallons of manure each year.

Tom also farms 1,600 acres of corn and soybeans and over 90% of it is no-tilled. He started no-tilling corn into bean stubble in 1980 and has been no-tilling beans for 5 years.

All Fall Application

He applies manure to 300 acres of corn stalks every fall. "I don't apply any manure in the spring," he says. "It's all applied in the fall or in the winter if we have to."

The one exception is summertime application on setaside acres. "I usually plant oats on setaside ground in the spring," Vittetoe says. "Then I pay my $10 to $12 per acre delayed clipping fee and let it start to rot. That's when I'll inject manure. The oats grow back at 35 to 40-bu. per acre and really suck up the nitrogen."

After the crop dies, it puts nitrogen back in the soil for corn or beans the next year. Oats also loosen the soil and produce a natural toxin that suppresses weeds.

Vittetoe has his crop rotation set up so he applies manure to each of his fields every 3 or 4

MEAN NUMBER OF COCOONS PRODUCED BY FIVE EARTHWORMS IN 3 MONTHS

Food Supply	A. chlorotica	L. castaneus
Fodder	0.8	9.4
Oat straw	1.4	12.0
Cattle manure	12.4	73.2
Sheep manure	14.0	76.0

—Edwards And Lofty

years. After knifing in the manure, he comes back with a finishing disk to close the knife tracks. "If you don't get those tracks closed, you could have real erosion problems," Vittetoe says.

Vittetoe estimates application takes 250 hours per year at a fuel cost of $1 per gallon and consumption of 28-gal. per hour. Labor runs around $10 per hour with two people involved full-

DON'T GET IT ON TOO THICK. Spread manure too thick on your fields and you'll cut earthworms off from their oxygen supply, say Univ. of Wisconsin entomologists. Lagoon aeration can also be a big concern.

FALL APPLIED MANURE WORKS BEST. Tom Vittetoe applies manure in each of his fields in the fall or winter every 3 or 4 years. After knifing in manure, he relies on a finishing disk to close up the knife tracks.

time and 1 half-time person.

Insect Control Needed

"If you use manure, you will have problems with seedcorn maggots and beetles," Vittetoe says. "A planter box treatment of Agrox DL will take care of the maggots. We have used Force insecticide on the beetles, but it has low water solubility.

"We're not disturbing life in the soil at all," he says. "The earthworms and all the rest are still there."

Quincy, Ill., ridge tiller Jim Callahan and his son, Randy, farm 700 acres of corn, soybeans and wheat as well as feeding out 2,000 hogs a year.

By applying manure to 160 acres, Callahan has maintained potassium and phosphorus levels without having to add commercial fertilizer. And, he'll point out, there are more nightcrawler holes than ever in that ground.

The Callahans have a custom built Calumet honey wagon. "Most have 90-in. wheel spacing while tractor tires are set at 60-in.," Callahan says. "That means you have to take time to move tractor tires when you want to apply manure."

There is a 120-in. spacing on their spreader, so it matches the tractor width.

Sources

Biology Of Earthworms, C.A. Edwards and J.R. Lofty, Chapman and Hall, 1977.

"If You Want More Earthworms," *Ridge Till Hotline,* April, 1993.

"No-Till Manure Management Means Subsurface Injection," *No-Till Farmer,* Mid-March, 1994.

Rotational Systems: The Key To Successful No-Till, Dwayne L. Beck, Dakota Lakes Research Farm, 1992, Zero Till Workshop, Manitoba-North Dakota Zero Till Farmers Assn.

Planning Pesticide Applications

EARTHWORMS HAVE the ability to metabolize and excrete low doses of carbofuran, an acutely toxic insecticide/nematicide.

FMC Corp., which manufactures Furadan, a carbofuran product, maintains worms below the first few inches of soil at application are not affected by Furadan.

Since the chemical does not have a long residual effect, treating a crop with Furadan may leave more than 80% of worms in the field unharmed.

"Earthworm numbers were cut back when we used Furadan on corn borers, but they are coming back now," says Nick Moore, who farms 700 acres under no-till and ridge till near Shabbona, Ill.

Early Bird Gets Worm, Dies

However, it was the impact of carbofuran on birds, partly by way of earthworms, that led to a historic agreement between FMC and the Environmental Protection Agency in 1991. The company consented to phase out granular Furadan from the marketplace within 3 years.

This was because EPA determined that consuming a single carbofuran granule could kill a small bird. Birds also may die after eating an earthworm contaminated with carbofuran.

In Virginia, the deaths of 200 songbirds in one corn field were traced to carbofuran application. This marked the first time the federal agency imposed restrictions on use of an agricultural chemical strictly because of its danger to wildlife. FMC has replaced granular Furadan with a liquid formulation that can be applied with a custom delivery system.

Worm Whippers

Carbofuran belongs to a class of insecticides and fungicides called carbamates. The majority of these products are toxic to

MOST HERBICIDES ARE HARMLESS TO EARTHWORMS. *Very few of today's herbicides are directly toxic to earthworms. However, application of herbicides that remove weeds during the early spring growth can actually end up depriving earthworms of food supplies.*

earthworms, according to Clive Edwards of Ohio State Univ., and Patrick Bohlen of the Institute for Ecosystem Studies in Millbrook, N.Y.

Other chemicals found in this category that have proved highly toxic to earthworms include aminocarb, methiocarb, oxamyl, promecarb, aldicarb, bufencarb, carbaryl, methomyl, propoxur and thiofanox. *(Check pages 93 to 100 to see how various pesticides measure up for earthworm toxicity and other key concerns).*

Carbamate fungicides, such as carbendazim and benomyl, have been shown to be moderately to highly toxic to earthworms. Broad-spectrum fumigants, applied to the soil to control fungi and nematodes, kill even earthworms that live in deep burrows. Contact nematicides also are very toxic to worms.

Many natural or synthetic pyrethroid insecticides have been tested, but none have proved deadly to earthworms, Edwards and Bohlen report.

Low Impact Insecticides

Most organophosphate insecticides are not very toxic to earthworms. Of all the members of this class, Dyfonate, Counter and Thimet seem to have the most effect on earthworm populations.

Herbicides Harmless

Very few modern herbicides are directly toxic to earthworms, although by removing weeds in early growth they can often deprive worms of some of their vauable food supply.

Some classes of crop chemicals, such as the triazine herbicides, may have moderate impact

PESTICIDES VS. WORMS. Bill Edwards, an Agricultural Research Service scientist at the North Appalachian Experimental Watershed in Coshocton, Ohio, had done extensive studies with herbicides and worm burrows.

Above, note length of worm hole. Stakes at upper right indicate holes used for pesticide runoff research. Bottom right photo shows canopy built to keep worms and worm holes from drying out. Research indicates no major environmental concern from chemicals moving through fields via worm holes.

on earthworm populations.

Inorganic chemicals, such as lead arsenate and copper sulfate, mainly used as pesticides before World War II, are moderately toxic to earthworms. The main threat to worm populations comes from long-term buildup of these chemicals in the soil.

The organochlorine insecticides, used extensively from the 1950 to the 1970s, are generally not very toxic to earthworms. Exceptions include the highly toxic chlordane and endrin and moderately toxic endosulfan and isobenzan.

Reducing Worm Impact

How much of an effect can insecticides have on earthworm populations and how fast? Ed Berry, entomologist at the National Soil Tilth Laboratory in

Ames, Iowa, conducted a series of experiments with insecticides applied at planting time.

It took 10 minutes for Furadan 15G applied directly on nightcrawlers to cause a noticeable response (which Berry defined as loss of coordination leading to death). Counter 15G ran a distant second, taking 3 hours to cause a noticeable response.

The worms responded to Thimet 20G after 5 hours, but there was no response to Lorsban 15G even after 5 hours.

Another Midwestern university study looked at average surface kill on earthworm plots treated at planting time with granular insecticides.

Based on the field trial results, standard rates of Furadan 15G will kill about 5,000 worms per acre. Dyfonate II 20G will kill 670, Thimet 20G, 580, Counter 15G, 500 and Lorsban 15G, 70 worms per acre.

However, different worm species may respond even to highly toxic materials in different ways, Edwards and Bohlen point out.

Carbofuran seems to be more toxic to Lumbricis terrestris than Eisenia foetida, while aldicarb is more toxic to E. foetida than L. terrestris. Also, differences in soil pH and organic matter content can influence insecticide impact on earthworms.

There also is evidence, based on studies in New Zealand, England and Nigeria, that earth-worms will avoid surface residue and soil layers that have been contaminated by pesticides. The worms stay below the surface and wait until chemical concentrations drop below toxic levels.

One thing you can do to reduce loss of earthworms to insecticides is avoid broadcasting toxic products in the spring and fall when earthworms normally are active near the surface.

Also, applying granular insecticide in a 1-in. band about 2-in. below and 2-in. to one side of the seed at planting greatly reduces earthworm mortality even when highly toxic insecticides are used.

Sources

"All Insecticides Not Equal When It Comes To Earthworms," *No-Till Farmer,* May, 1993.

"Helpful Earthworms Can Rebound From Pesticides," *Delmarva Farmer,* March 22, 1994.

Earthworms And Crop Management, Eileen Kladivko, Purdue Univ. Cooperative Extension Service, 1993.

"Effectively Managing Those Highly Valuable Earthworms When No-Tilling," Eileen Kladivko, National No-Tillage Conference, St. Louis, Mo., Jan. 11, 1994 (available on audiocassette).

"EPA Grounds Carbofuran," Sonja Hillgren, *Farm Journal,* August, 1991.

"Flowable Furadan Fits No-Till," *No-Till Farmer,* May, 1993.

The Effects Of Toxic Chemicals On Earthworms, Clive A. Edwards and Patrick J. Bohlen, Ohio State Univ.

How 193 Chemicals Affect Earthworms

THE FOLLOWING is an analysis of the "Summary Of The Results Of Laboratory And Field Experiments" from *The Effects Of Toxic Chemicals On Earthworms.*

It was compiled by Clive A. Edwards of Ohio State Univ. and Patrick J. Bohlen, now at the Institute for Ecosystem Studies in Millbrook, N.Y.

One chemical not included in this extensive list is tefluthrin, the active ingredient in Force insecticide. There hasn't been a lot of research done on this relatively new product, but Edwards says like other synthetic pyrethroids, it can be considered harmless to worms.

Chemical Tested	Trade Names	Relative Toxicity	Comments
Inorganic Chemicals...			
Calcium arsenate		0	
Calcium cyanimide		NS	Probably non-toxic at normal exposure rates
Copper chloride		NS	Probably non-toxic at normal exposure rates
Copper oxychloride		0	Persistent, toxic at high exposure rates

---

Relative Toxicity Scale...

0 = Non-toxic
1 = Slightly Toxic
2 = Moderately Toxic
3 = Very Toxic
4 = Extremely Toxic
NS = Insufficient Evidence

Chemical Tested	Trade Name	Relative Toxicity	Comments
Copper sulfate		0	Persistent, toxic at high exposure rates
Lead arsenate		0	Persistent, toxic at high exposure rates
Sulfur		2	
Mercuric chloride		NS	Probably very toxic at high exposure rates
Potassium bromide		2	
Potassium Permanganate		0	Expels earthworms from soil
Sodium chlorate		0	

Biological Agents...

Enterobacterin		0	
Mowrah meal		0	Expels earthworms from soil
Rotenone		0	Expels earthworms from soil
Mustard		NS	Probably non-toxic at normal exposure rates

Aromatic And Organochlorine Insecticides...

Aldrin		0	Non-toxic except at high exposure rates
Aramite		NS	Probably non-toxic at normal exposure rates
BHC		0	Non-toxic except at high exposure rates
Chlordane		4	
DDE		NS	Probably non-toxic at normal exposure rates
DDT		0	
Dicofol		NS	Probably non-toxic at normal exposure rates
Dieldrin		0	
DNOC		1	
Endosulfan	Thiodan Phaser	2	
Endrin		4	
Heptachlor		4	
Isobenzan		3	Variable; very toxic in some field tests
Lindane		NS	Probably toxic at normal exposure rates
Napthalene		NS	Probably very toxic at normal exposure rates
1-Naphthol		NS	Probably very toxic at normal exposure rates
Telodrin		1	

Chemical Tested	Trade Name	Relative Toxicity	Comments
Tetradifon		0	
Toxaphene		2	

Organophosphate Insecticides...

Chemical Tested	Trade Name	Relative Toxicity	Comments
Acephate	Orthene, Payload	NS	Probably toxic at normal exposure rates
Azinphos methyl	Guthion	NS	Probably toxic at normal exposure rates
Bromophos		NS	Probably slightly toxic at normal exposure rates
Carbophenothion		NS	Probably slightly toxic at normal exposure rates
Chlorfenvinphos		1	
Chlormephos		NS	Probably non-toxic at normal exposure rates
Chlorpyrifos	Dursban Empire Equity Estate Lentrek Lock-On Pageant Tenure	3	
Chlorpyrifos-ethyl	Lorsban	NS	Probably non-toxic at normal exposure
Demeton-S-methyl		NS	Probably non-toxic at normal exposure rates
Dialifos		NS	Probably non-toxic at normal exposure rates
Diazinon	Knox Out Diazol	1	
Dimethoate	Cygon	1	
Disulfoton	Mocap Plus Disyston	1	
Ethoprophos	Mocap	3	
Ethyl-parathion		3	Extremely toxic at high exposure rates
Fenamiphos	Nevacer	1	
Fenitrothion		0	
Fensulfothion		1	Variable; probably non-toxic
Fonofos	Dyfonate	2	
Formothion		1	
Isazophos		3	
Isofenphos		1	
Leptophos		NS	Probably slightly toxic at normal exposure rates
Malathion	Cythion	1	

Chemical Tested	Trade Name	Relative Toxicity	Comments
Menazon		0	
Methamidophos		NS	Probably slightly toxic at normal exposure rates
Methaphenamiphos	Monitor	NS	Probably slightly toxic at normal exposure rates
Methidathion		NS	Probably toxic at normal exposure rates
Methyl-parathion	Penn Cap	NS	Probably non-toxic at normal exposure rates
Monocrotophos		0	
Paraoxon		NS	Probably non-toxic at normal exposure rates
Phorate	Thimet	4	
Phosalone		1	
Phosphamidon	Swat	NS	Probably non-toxic at normal exposure rates
Terbufos	Counter	2	
Tetrachlorvinphos		0	
Thionazin		0	
Triazophos		NS	Probably non-toxic at normal exposure rates
Trichloronate		0	
Trichlorfon		1	

Carbamate Insecticides...

Chemical Tested	Trade Name	Relative Toxicity	Comments
Aldicarb	Temik	4	
Aminocarb		NS	Probably toxic at normal exposure rates
Bendiocarb	Rotate Ficam	4	
Bufencarb		3	
Carbaryl	Sevin	4	
Carbofuran	Furadan	4	
Ethiofencarb		NS	Probably non-toxic at normal exposure rates
Methiocarb	Measural	3	
Methomyl	Lannate	4	
Oxamyl	Vydate	2	
Promecarb		NS	Probably non-toxic at normal exposure rates
Propoxur	Baygon	3	
Thiofanox		3	

Synthetic Pyrethroids And Chitin Inhibitors...

Chemical Tested	Trade Name	Relative Toxicity	Comments
Alphamethrin	Ammo Cymbush	NS	Probably non-toxic at normal exposure rates

Chemical Tested	Trade Name	Relative Toxicity	Comments
Cypermethrin	Demon Ammo Prevail	1	
Diflubenzuron	Dimilin	0	
Fenvalerate	Pydrin	0	
Permethrin	Ambush Pounce Torpedo	NS	Probably non-toxic at normal exposure rates

Soil Fumigants And Nematicides...

Chloropicrin	Chlor-O-Pic	4	
Dazomet	Basamid	2	
DBCP		0	
Ethylene dibromide		NS	Probably very toxic at normal exposure rates
1,2-Dichloropropane		NS	Probably non-toxic at normal exposure rates
Dichloropropane-dichloropropene		4	
Formaldehyde		2	Expels earthworms from soil
Metham sodium		NS	Probably very toxic at normal exposure rates
Methyl bromide	Terr-O-Gas Meth-O-Gas		Probably very toxic at normal exposure rates

Fungicides...

2-Aminobutane		NS	Probably non-toxic at normal exposure rates
Aniyaline		1	
Benomyl	Benlate Tersan	4	
Buprimate		NS	Probably non-toxic at normal exposure rates
Captafol		0	
Captan		1	
Carbendazim		3	
Chlorthalonil	Bravo Daconil	2	
Dichloran		0	
Dinocap		NS	Probably non-toxic at normal exposure rates
Ethazole		NS	Probably non-toxic at normal exposure rates
Fenaminosulf		NS	Probably non-toxic at normal exposure rates
Fenarimol		NS	Probably slightly toxic at

Chemical Tested	Trade Name	Relative Toxicity	Comments
Folpet		NS	normal exposure rates Probably non-toxic at normal exposure rates
Fuberidazole		NS	Probably non-toxic at normal exposure rates
Imazalil	Fecundal Fungaflor	NS	Probably toxic at normal exposure rates
Mancozeb	Manzate Penncozeb Dithane Fore	0	
Maneb		NS	Probably non-toxic at normal exposure rates
Propiconazole		NS	Probably toxic at normal exposure rates
Quintozene		NS	Probably non-toxic at normal exposure rates
Thiabendazole		3	
Thiophanate-methyl		3	
Thiram		0	
Triadimefon		0	
Triforine		NS	Probably non-toxic at normal exposure rates
Ziram		NS	Probably non-toxic at normal exposure rates

Herbicides...

Chemical Tested	Trade Name	Relative Toxicity	Comments
Aminotriazole		0	
Asulam		NS	Probably non-toxic at normal exposure rates
Atrazine	Aatrex	0	
Aziprotryne		NS	Probably non-toxic at normal exposure rates
Bromacil	Hyvar	0	
Chlormequat chloride		NS	Probably non-toxic at normal exposure rates
Chloroacetamide		4	
Chlorpropham		1	
Chorthiamid		2	
Chlortoluron		NS	Probably non-toxic even at high exposure rates
Cyanazine	Bladex	0	
Cycloate		NS	Probably non-toxic at normal exposure rates
2,4-D	Butoxone Weedone	1	Non-toxic, except at high exposure rates

Chemical Tested	Trade Name	Relative Toxicity	Comments
Dalapon		0	
Di-allate		NS	Probably non-toxic at normal exposure rates
Dicamba	Banvel		
	Trooper	NS	Probably non-toxic at normal exposure rates
Dinoseb		2	
Diphenamid		NS	Probably slightly toxic at normal exposure rates
Diquat		NS	Probably non-toxic at normal exposure rates
Diuron	Karmex	0	
Endothal		NS	Probably non-toxic at normal exposure rates
Glyphosate	Roundup	0	
Hexazinone	Velpar	NS	Probably non-toxic at normal exposure rates
Lenacil		NS	Probably non-toxic at normal exposure rates
Linuron	Lorox Linuron	0	
Maleic hydrazide		0	
MCPA		0	
MCPB		NS	Probably non-toxic at normal exposure rates
Mecoprop		NS	Probably non-toxic at normal exposure rates
Methabenzothiazuron		0	
Metribuzin	Lexone Sencor	NS	Probably non-toxic at normal exposure rates
Monolinuron		NS	Probably non-toxic at normal exposure rates
Monuron		0	
Nitrofen		NS	Probably non-toxic at normal exposure rates
Oxadiazon		NS	Probably non-toxic at normal exposure rates
Paraquat	Gramoxone Extra	0	
Pendimethalin	Prowl Pendulum	NS	Probably toxic at normal exposure rates
Pentachlorophenol		4	
Phenmedipham		NS	Probably non-toxic at normal exposure rates
Prometryn		NS	Probably non-toxic at normal exposure rates
Propazine		NS	Probably non-toxic at

Chemical Tested	Trade Name	Relative Toxicity	Comments
			normal exposure rates
Propham		NS	Probably slightly toxic at normal exposure rates
Pyrazone		1	
Sesone		NS	Probably non-toxic at normal exposure rates
Simazine	Princep	2	
Sodium trichloroacetate		0	
2,4,5-T		NS	Probably non-toxic at normal exposure rates
2,3,6-TBA		NS	Probably non-toxic at normal exposure rates
Terbacil	Sinbar	0	
Trichlopyr	Turflon	NS	Probably non-toxic at normal exposure rates
Trichloroacetic acid		1	
Tri-allate		0	
Trifluralin	Treflan Tri-4 Trust	0	

Other Organic Compounds...

Chemical Tested	Trade Name	Relative Toxicity	Comments
3-Chlorophenol		4	
Dichloroaniline		NS	Probably toxic at normal exposure rates
3,4-Dichlorophenol		4	
Dimethyl phthalate		NS	Probably non-toxic at normal exposure rates
Fluorene		NS	Probably toxic at normal exposure rates
Hexoestrol		NS	Probably toxic at normal exposure rates
p-Nitrophenol		4	
N-Nitrosodiphenylamine		NS	Probably toxic at normal exposure rates
Dioxin		2	
Nitrobenzene		NS	Probably toxic at normal exposure ratesd
Phenol		NS	Probably toxic at normal exposure rates
2,3,4,5-Tetrachlorophenol		3	
1,2,3-Trichlorobenzene		NS	Probably toxic at normal exposure rates
1,2,4-Trichlorobenzen		NS	Probably toxic at normal exposure rates
2,4,5-Trichlorophenol		4	
2,4,6-Trichlorophenol		4	

—Clive A. Edwards and Patrick J. Bohlen

Cover Crops, Crop Rotations

"WE LIKE TO FEED earthworms with plenty of organic matter from forages, cover crops and manure while minimizing soil disturbance," says Ben Stinner, an Ohio State Univ. agronomist. That pretty much sums up the advantages of seeding cover crops in double-crop or multi-crop rotations.

Cover crops like rye, clover, wheat or hairy vetch build soil organic matter which benefits earthworms. They also can reduce the need for insecticides, herbicides and commercial fertilizer which may harm worms.

"Ground cover in the form of crop residue, rye or vetch stops erosion and makes a feeding area for earthworms," says Wayne Livengood, Terre Haute, Ind., farmer. "These worms aerate the soil, make it easier for more rain to soak in and less to run off."

Thanks to less tillage, cover crops and worms, Livengood has eliminated commercial fertilizer except for liquid nitrogen.

Rotation Pays Off

Rotation boosts earthworm populations as well as crop yields. A survey of fields in Indiana and Illinois showed very few worms in continuous corn, regardless of whether the crop was conventionally tilled or no-tilled.

Continuous soybean plots had higher earthworm populations than continuous corn. "Earthworms generally prefer legumes as a food source over grasses," says Eileen Kladivko, Purdue Univ. soil scientist. "This is probably the main reason for the higher populations in the beans."

The continuous corn plots also were treated with rootworm insecticide and anhydrous ammonia, Kladivko points out. The inclusion of a legume in the cropping pattern plus the elimination of insecticide means more earthworms in a corn/bean rotation.

Dick Thompson doesn't even consider corn followed by soybeans a true rotation. On his Boone, Iowa, farm, Thompson follows a 6-year rotation of corn, soybeans, oats and 3 years of pasture that includes a mixture of alfalfa, red clover and other forage crops that take full advantage of growing conditions.

In 1989, the National Soil Tilth Laboratory at Ames, Iowa, initiated a study of Thompson's system compared to a neighbor's conventionally tilled field. The researchers found, among other things, more earthworms in Thompson's fields.

Doug Karlen, the project leader, attributes this improvement to reduced tillage and addition of carbon, a food for earthworms, through manure application and planting of cover crops and rotation of row crops with oats and hay.

Soil Compaction Solution

Rotating no-till cotton with corn or a small grains cover crop can solve surface compaction problems and encourage earthworm growth in the red clay loam soils of the Tennessee River Valley, says Ken Rogers, Soil Conservation Service agronomist in Auburn, Ala.

Unlike most clay soils, red clay doesn't shrink when it's dry or swell when wet. As a rule, organic matter runs less than 1%. Worst of all, these soils are prone to surface compaction in top 2 to 3-in.

"Too many years of continuous cotton and tillage have resulted in poor soil structure and low organic matter," Rogers says.

Corn, however, has a strong fibrous root system which grows better in compacted soils than cotton roots. Planting wheat or rye ahead of cotton in the fall will also loosen up the soil.

Crop residue from corn or small grains not only breaks up compaction, but provides food for earthworms. In turn, more worm burrows further improve water filtration and provide even more pathways for root growth.

No-till cotton and alfalfa studies in California have shown similar results.

An Unexpected Advantage

Bill Rohrs, coordinator of the Conservation Action Project in northwestern Ohio, discovered yet another cover crop benefit while talking to two Ohio farmers at the 1993 National No-Tillage Conference in Indianapolis, Ind.

They told Rohrs that worms removed most of the residue from their corn fields in 6 months.

"Their praise of the critters could not be surpassed," Rohrs says. "But they were worried about how they would meet the 30% residue requirement for conservation compliance after a 180-bu. per acre crop—a wonderful problem to have!"

WORMS INFLUENCE AVAILABLE MOISTURE*

(With and without worms in New Zealand pasture)

Sampling Time	Pasture With Lumbricids**	Pasture Without Lumbricids**
June	8.5 cm.	7.9 cm.
August	8.4	6.6
November	3.4	2.5
February	6.4	5.6

Available moisture measured in centimeters at a depth of 0 to 30 centimeters.

** *Based on a count of 1,150 worms per square meter.*

Seasons in New Zealand are opposite of North America.

—K.E. Lee

These farmers' experience are backed up by Univ. of Wisconsin no-till corn research. Worms reduced residue cover from 30% to as low as 15% at planting time, says Andy Gallagher.

If earthworms consume too much of the residue on top of your field, Rohrs suggests seeding rye as a cover crop.

"Once used in corn so harvesting could continue in wet, smeary conditions, it is time to consider rye seedings for other reasons," Rohrs says. "Rye adds tremendous root structure, sometimes growing 5 to 6-ft. deep." It's as easy to kill as clover, whether followed by corn or soybeans.

Bill Becker, director of the Central Illinois Research Farm at Springfield, Ill., favors hairy vetch. He suggests flying it onto soybeans in August and then interseeding oats after soybean harvest.

"The oats tie up any excess nitrogen before winterkill," Becker says. "This forces the vetch to manufacture more nitrogen on its own."

Disking the vetch lightly before planting corn increases soil nitrogen and earthworm populations even more. Seed rye or oats after corn harvest to tie up nitrogen and provide fall cover, Becker advises. Spraying corn stalks with 30 units of 28% nitrogen reduces the carbon-nitrogen ratio and improves palatibility to earthworms.

Exotic Cover Crops

In addition to familiar grasses and legumes, there are several new cover crop options.

Research at Ontario's Univ. of Guelph show Austrian peas can provide 235-lbs. per acre of available nitrogen for any following no-till corn crop.

Keith Bakken of Alden, Minn., was especially interested in finding an alternative to chemical weed control when growing amaranth. The specialty crop is related to pigweed and will not tolerate any herbicides.

Medic, an annual legume imported from Australia, provided Bakken—who also ridge tills corn and soybeans—with that alternative. Related to alfalfa, medic varieties flower, set seed and die within a single season.

If planted with row crops in the spring, medic will be dead by August 1, leaving behind a thick ground cover that smothers weeds and protects the soil. As a legume, it also fixes up to 100-lbs. per acre of soil nitrogen.

Since Bakken hasn't used any insecticide in 5 years, it's easy for him to put medic seed in the planter's insecticide box and interseed it with his row crops.

"Anything you can add to give you more residue on the field benefits earthworms," says Maltie Hawkinson, a Center, Mo., wheat and soybean farmer who often uses wheat as a winter cover crop. He'll burn the wheat down in the spring before planting soybeans.

"With a cover crop, the ground doesn't cool off as fast in the fall," he explains. "This means the worms will continue working longer. When it gets too cold for them, they'll move deeper in the soil." A cover crop means more worm food, too.

Sources

Earthworms And Crop Management, Eileen Kladivko, Purdue Univ. Extension Service.

Earthworms: Their Ecology And Relationships With Soils And Land Use., K.E. Lee, Academic, Press, 1985.

Growing Corn Without Fertilizer? Wayne F. Livengood, January, 1994.

"How Can I Increase Earthworm Numbers In No-Tilled Fields?" *No-Till Farmer,* May, 1993.

"Hungry Crawlers: Too Much Of A Good Thing?" *No-Till Farmer,* Mid-March, 1993.

"Legume Provides Herbicide Alternative," *Ridge Till Hotline,* May, 1994.

"Let Austrian Winter Peas Smother Out," *No-Till Farmer,* May, 1992.

"Lowly Creatures Rank High In Soil Building Benefits," *Iowa Farm Bureau Spokesman,* April 21, 1990.

"Systems Approach Really Works!" Frank Lessiter, *Ridge Till Hotline,* June, 1994.

Thompson On-Farm Research With Wallace Institute, 1993 Report, April 13, 1994.

Earthworm Questions And Answers

HERE ARE A NUMBER of typical questions which farmers regularly ask about earthworms and their environment.

Q: *If I switch to earthworm friendly practices, how long will it take for worms to come back to my fields in large numbers?*

A: The time for earthworm populations to stabilize under favorable conditions ranges from 3 to 8 years.

Q: *What effects do tillage and agricultural chemicals really have on earthworm populations in my fields?*

A: Tillage that leaves little or no residue on the surface, such as moldboard plowing, creates unfavorable conditions for earthworms. So does excessive fall tillage of any kind.

A quick hard freeze under these conditions doesn't give the worms time to migrate deep into the soil for protection against the cold.

Shallow tillage is better than deep tillage and conservation tillage systems (which leave at least 30% of crop residue on the soil surface) create the most favorable earthworm environments.

Most insecticides, as currently used, are relatively non-toxic to earthworms. Some insecticides are highly toxic, including some soil applied insecticides.

Triazine herbicides (such as atrazine) are slightly toxic to worms, but many others are non-toxic.

High concentrations of ammonium-based fertilizers are toxic to earthworms. Long-term exposure to these fertilizers injures earthworm health.

Q: How do I know how many earthworms I have in my fields

A: Counting worms takes a lot of time and work. But it's relatively simple to tell if there are earthworms present in your fields. Look for residue mounds (huts or middens), trails, actual worms around the corn crowns, castings on the surface and earthworms of various sizes counted in shovelfuls of soil.

Q: Will improved water infiltration speed movement of chemicals away from the root zone and into the groundwater?

A: Bill Edwards, chief researcher at the USDA North Appalachian Experimental Watershed, Coshocton, Ohio, says vertical nightcrawler burrows can provide flow paths to carry nitrates and other agricultural chemicals down into the soil, especially in reduced tillage situations where farmers don't disturb burrows by working the soil. But just because the chemicals may go down a few feet doesn't mean they will end up in groundwater.

During a typical summer at Coshocton, less than 4% of rainwater that infiltrates the soil will flow into deep nightcrawler burrows. Twelve storms strong enough to put water in nightcrawler holes may send 1/2 to 1-lb. of nitrates down into the soil.

But temporary horizontal burrows created by other earthworm species in the first foot of topsoil improves soil water holding capacity—which means less surface runoff to carry fertilizer and pesticides into lakes and streams.

Worm burrows also improve soil aeration, which means more microbial activity to break down chemicals.

A light rain will carry atrazine and alachlor herbicide down vertical burrows just far enough to mix with and bind to organic matter. If a major storm (3 or more inches of rain in an hour) follows the light rain, not much more herbicide will move downward. If the field is hit by a heavy rain first, more of the herbicides filter down through the soil.

This means there's a chance of chemicals contaminating groundwater by running down earthworm holes, but the odds are against conditions being just right for this to happen.

In addition, research at Ohio State Univ. and the USDA Hydrology Laboratory in Beltsville, Md., indicates the material with which nightcrawlers line their burrows absorbs herbicides and slows their flow.

Q: How can I minimize

the risks of groundwater contamination?

A: Injecting fertilizer below the soil surface prevents nitrates washing down wormholes. Switching to less water-soluble and less persistent herbicides can reduce the potential for groundwater contamination from this source.

Q: What happens to earthworms during flooding? Do they drown?

A: Research from several areas of Iowa that were flooded in 1993 shows earthworms are not affected when rivers or creeks overflow for short periods of time. However, prolonged flooding wipes out worm populations. In general, earthworms do not drown if provided with an oxygen source and the temperature of the water is not extremely hot or cold.

Q: What about earthworms under flood irrigation?

A: When earthworms wriggle toward water flowing through an irrigation furrow, their tiny tunnels drain water down and away from the trench. That leaves part of the field high and dry, says Thomas J. Trout, agricultural engineer at the USDA Soil and Water Management Research

Unit at Kimberly, Idaho.

Water infiltration rates may increase as much as 20% in worm-riddled furrows. That means you may have to increase furrow flows to move water across the field—which could increase soil erosion.

Adding low concentrations of aqueous ammonia to irrigation water will drive worms away. Other options include switching to sprinkler irrigation or reduced tillage.

Q: Do earthworms carry parasites and disease?

A: Yes, but most are harmful only to worms or birds and other creatures which prey on them. Nematodes which infest the throats of birds are very common.

Earthworms may carry parasites which invade the lungs, kidneys and stomachs of pigs. In other countries, they may transmit hoof and mouth disease to cattle if chopped plant material infected with the virus is incorporated into the soil.

On the other hand, earthworms seem to reduce populations of salmonella bacteria in sewage sludge applied to fields as fertilizer.

Sources

"Can Chemicals Worm Their Way Into Groundwater?" Syl Marking,

PSYCHIC EARTHWORMS?

Some researchers have observed a very strange aspect of earthworm behavior. Earthworms under carefully controlled laboratory conditions follow biological cycles similar to those found in the field.

Andre Kretzschmar, a scientist from Montfavet, France, gathered worms to study their response to soil compaction. He collected them from a pasture early in the year when the soil was quite dry and the worms were very active.

They were maintained in soil compacted columns under carefully monitored temperature, lighting and other conditions. "All 175 replicates burrowed nicely and casted actively until the beginning of June—then the activity ceased," Kretzschmar notes.

"More surprisingly, fresh casts reappeared in all of the columns between August 25 and Oct. 9. Exactly at the same time their friends in the field were doing the same job!"

Soybean Digest, Mid-March, 1990.

"Characterizing Macropores That Affect Infiltration Into Non-Tilled Soil," W.M. Edwards, L.D. Norton and C.E. Redmond, *Soil Science Society Of America Journal,* Volume 52, No. 2, March-April, 1988.

"Contribution Of Macroporosity To Infiltration Into A Continuous Corn No-Tilled Watershed: Implications For Contaminant Movement," W.M. Edwards, M.J. Shipitalo and L.D. Norton, *Journal Of Contaminant Hydrology,* 1988.

"Do Earthworms Affect Groundwater?" Ben Hardin and Don Comis, *Agricultural Research,* April 1991.

Earthworms: Their Ecology And Relationships With Soils And Land Use, K.E. Lee, Academic Press, 1985.

"Effect Of Formulation And Tillage Practice On Volatilization Of Atrazine And Alachlor," Brian J. Wienhold and Timothy J. Gish, *Journal Of Environmental Quality,* April, 1994.

"Irrigation Versus The Earthworm," Julie Corliss, *Agricultural Research,* April 1991.

Leopold Letter, Vol. 5, No. 4, Winter, 1993, The Leopold Center for Sustainable Agriculture, 126 Soil Tilth Building, Iowa State University, Ames, IA 50011-3120.

"Sorption And Retention Of Herbicides In Vertically Oriented Earthworm And Artificial Burrows," R.C. Stehouwer, W.A. Dick and S.J. Traina, *Journal Of Environmental Quality,* April, 1994.

"Why Earthworms Are A Big, Big No-Till Bonus," Bill Edwards, 1993 National No-Tillage Conference, Indianapolis, Ind.

Where To Find Even More Earthworm Information

WE HOPE THAT we have shown you in this book how earthworms can really help your farming operation, regardless of the tillage or croping system which you use.

If you want to learn still more about earthworms, the people listed in this chapter can answer many of your questions and provide much additional information.

John Alexander
12005 Boundary Rd.
Fountain City, IN 47341

Ernie Behn
1309 Quill Ave.
Boone, IA 50036
515/432-1569

Edwin C. Berry
National Soil Tilth
Laboratory
2150 Pammel Drive
Ames, IA 50011-4420
515/294-5723

John Buckerfield
CSIRO Division of Soils
PMB 2, Glen Osmond
South Australia 5064

Kevin R. Butt
Biosystems Group,
Faculty of Technology
Open University
Milton Keynes
England MK8 6AA
908-653548

Bill Edwards
North Appalachian
Experimental Watershed
Agricultural Research
Service
United States
Department of Agriculture
P.O. Box 478,
Coshocton, OH 438
614/545-6349

Clive Edwards
Department of Entomology
Ohio State University
103 Botany & Zoology
Building
1735 Neil Ave.
Columbus, OH 43210
614/292-3786

Diana W. Freckman
Natural Resource Ecology
Laboratory
College of Natural
Resources
Colorado State University
Fort Collins, CO 80523
303/491-1982
Fax: 303/491-1965

Maltie Hawkinson
R.R. 2,
Box 2706,
Center, MO 63436
314/267-3466

P.F. Hendrix
Institute of Ecology
University of Georgia
Athens, GA 30602

Jim Kinsella
R.R. 2
Box 138-A
Lexington, IL 61753
309/365-2111

Eileen Kladivko
Department of Agronomy
Purdue University
West Lafayette, IN 47907
317/494-6372

Dave Larson
AgriEnergy Resources
R.R. 2
Box 113
Princeton, IL 61356
815/872-1190
Fax: 815/872-1928

Ken E. Lee
CSIRO Division of Soils
PMB 2,
Glen Osmond
South Australia 5064

Leopold Center for
Sustainable Agriculture
126 Soil Tilth Building
Iowa State University
Ames, IA 50011-3120
515/294-3711

Dennis R. Linden
Soil and Water
Management Research
Unit,
Northern States Area
Agricultural Research
Service
St. Paul, MN 55108

Wayne Livengood
9666 Rio Grande Ave.
Terre Haute, IN 47805

Dennis Parkinson
Department of
Biological Sciences
University of Calgary
Calgary, Alberta T2N 1N4

Randy Quilliam
2715 Canal Rd. NE
Millersport, OH 43046
614/862-4781

John Reynolds
Sir Sandford Fleming
College
P.O. Box 8000
Lindsay, Ontario
Canada K9V 5E6
705/324-9144
Fax: 705/878-9312

William F. Rohrs
Conservation Action Project
197-1B Island Park Ave.
Defiance, OH 5389
419/784-5389

John Satchell
Draw Well,
Lyth, Kendal
Cumbria LA8 8DF England

Art Stieren
R.R. 2
Box 113
West Point, NE

Alan Tomlin
London Research Centre
Agriculture Canada
1391 Sandford St.
London, Ontario N5V 4T3
519/645-4452
Fax: 519/645-5476

Dick and Sharon Thompson
Practical Farmers Of Iowa
2035 190th St.
Boone, IA 50036
515/432-1560

Tom Vittetoe
2460 275th St.
Washington, IA 52353

WHAT DID ROSS PEROT REALLY HEAR?

In a memorable quote, Ross Perot identified "a giant sucking sound" as U.S. jobs were being drawn to Mexico. But a ridge tiller from Farmer City, IIll, heard something else. Once after a big rainstorm, Dave Beckenholdt, took a walk in his fields.

"It was real quiet—not a sound to be heard," he recalls. "Then I stomped my foot on the ground and I heard a big sucking noise as all the earthworms that had come to the top of the soil retreated back down into their burrows."

More Quotable Quotes...

"Earthworms come to the surface and eat residue, leaving a rich end product. They also incorporate my phosphorus and potassium with ridge till..."

—*Loren Schuett,*
Holstein, Iowa

"Earthworms are nature's tillers. They're cheap and do an excellent job..."

—*Jim Kinsella,*
Lexington, Ill.

"With no-till, you don't disturb the earthworm holes. These holes hold the water and channel it deeper into the soil. No-till reduces runoff tremendously..."

—*Ken Rutz,*
LaPorte County, Ind.

"We are definitely improving both structure and organic matter in our soils..."

—*Randy Quilliam,*
Millersport, Ohio

"If an earthworm gets in front of an anhydrous shank, that sucker's going to die! But we drown more earthworms fishing than we hit with anhydrous ammonia shanks..."

—*George Rehm,*
Univ. of Minnesota
St. Paul, Minn.

"I don't sidedress nitrogen in our corn real early in the morning or real late at night. Those are the times when earthworms are up in the danger zone..."

—*Greg DeMott,*
Ashley, Mich.

"I have always felt if the soil was loose with a lot of air in it and if it also had a lot of earthworms...then a split application of 28% liquid nitrogen over the top would have to be easier on the earthworm population..."

—*Mike Wurmnest,*
Deer Creek, Ill.

"Burrows left by worms are nature's own system for collecting rainwater in the soil where plant roots can suck it up. This gives less water runoff to strip away valuable topsoil and desposit chemicals in waterways..."

—*Edwin Berry,*
National Soil Tilth Lab,
Ames, Iowa